Lecture Notes in Civil Engineering

Volume 114

Lecture Notes in Civil Engineering (LNCE) publishes the latest developments in Civil Engineering - quickly, informally and in top quality. Though original research reported in proceedings and post-proceedings represents the core of LNCE, edited volumes of exceptionally high quality and interest may also be considered for publication. Volumes published in LNCE embrace all aspects and subfields of, as well as new challenges in, Civil Engineering. Topics in the series include:

- Construction and Structural Mechanics
- Building Materials
- Concrete, Steel and Timber Structures
- Geotechnical Engineering
- Earthquake Engineering
- Coastal Engineering
- Ocean and Offshore Engineering; Ships and Floating Structures
- Hydraulics, Hydrology and Water Resources Engineering
- Environmental Engineering and Sustainability
- Structural Health and Monitoring
- Surveying and Geographical Information Systems
- Indoor Environments
- Transportation and Traffic
- Risk Analysis
- Safety and Security

To submit a proposal or request further information, please contact the appropriate Springer Editor:

- Pierpaolo Riva at pierpaolo.riva@springer.com (Europe and Americas);
- Swati Meherishi at swati.meherishi@springer.com (Asia - except China, and Australia, New Zealand);
- Wayne Hu at wayne.hu@springer.com (China).

All books in the series now indexed by Scopus and EI Compendex database!

More information about this series at http://www.springer.com/series/15087

K. Thirumaran · G. Balaji · N. Devi Prasad
Editors

Sustainable Urban Architecture

Select Proceedings of VALUE 2020

Springer

Editors
K. Thirumaran
Department of Architecture
National Institute of Technology Tiruchirappalli
Tiruchirappalli, Tamil Nadu, India

G. Balaji
Thiagarajar College of Engineering
Madurai, Tamil Nadu, India

N. Devi Prasad
School of Architecture
Vellore Institute of Technology
Vellore, Tamil Nadu, India

ISSN 2366-2557 ISSN 2366-2565 (electronic)
Lecture Notes in Civil Engineering
ISBN 978-981-15-9584-4 ISBN 978-981-15-9585-1 (eBook)
https://doi.org/10.1007/978-981-15-9585-1

This Springer imprint is published by the registered company Springer Nature Singapore Pte Ltd.
The registered company address is: 152 Beach Road, #21-01/04 Gateway East, Singapore 189721, Singapore

Organization

First International Conference on Visionary Action Towards Liveable Urban Environments

Value 2020
20–21 February 2020

Organized by

School of Architecture, Vellore Institute of Technology, Vellore

Supported by

International Scientific Committee

Dr. Marc Aurel Schnabel
Dean
School of Architecture
Victoria University of Wellington

Dr. Mohd Hamdan Bin Haji Ahmad
Dean
Faculty of Built Environment
Universiti Teknologi Malaysia

Prof. Ar. Hans-Christian Wilhelm
Senior Lecturer
School of Architecture
Victoria University of Wellington

Dr. Sandeep Agrawal
Director
Planning Program
Department of Earth and Atmospheric Sciences
University of Alberta

National Scientific Committee

Ar. Diwakar Chintala
Principal Architect
Studio Chintala
Bengaluru

Dr. Abraham George
Associate Professor
Department of Architecture and Regional Planning
IIT Kharagpur

Dr. V. Devadas
Professor
Department of Architecture And Planning
IIT Roorkee

Ms. Pallavi Kulkarni
Urban Designer and Planner
DPadeco Co. Ltd.

Technical Programme Committee

General Chair:
Prof. Devi Prasad. N, Dean, School of Architecture, VIT Vellore

Organizing Chair:

Prof. A. Madhumathi

Prof. Sharmila Jagadisan

Hospitality Chair:
Prof. Rajabhooshanam Arlene
Dr.Djamil Ben Guida
Prof. Jayprakash
Prof. Michael Karassowitsch
Prof. Kota Sandeep
Prof. Dinesh Raghavan
Prof. Meenakshi Pappu
Prof. Mohafiz Riyaz
Prof. Jayadev Nallari
Prof. Ankit Kumar
Prof. Bhaskar Jyoti Borgohain
Prof. Shreya Mukherjee
Prof. M. Dipika
Srinivasan. A, DEO, School of Architecture

Preface

The 1st International Conference on Visionary Action towards Liveable Urban Environments (VALUE 2020) was held on 20 and 21 February 2020 in Vellore Institute of Technology, Vellore, India. VALUE 2020 provided an academic platform for researchers to present their latest findings to demonstrate how liveability encompasses a wide range of critical factors relating to overall quality of life and well-being. As the world continues to urbanize, it is extremely crucial to link every aspect of urbanization with sustainability. The concept of 'liveability' has emerged alongside 'sustainability' as a keyword in public discourse and planning and has no universally agreed-upon definition. Liveability is intrinsically linked to the economy, social inclusion, health and well-being for some people, and for others, it implies urban green space intervention. Architects and urban planners have come to understand that improving the quality of life in cities is no longer a simple matter of changing physical entities, but also involves popular satisfaction with different urban attributes such as public transit, quality and safety of public spaces, recreational facilities, ease of access for all to basic goods, services and public amenities. Over the past few decades, the gap between theoretical knowledge and practice has been increasing especially in architecture and urban planning where there is an astonishing lack of critical perspectives. To understand the multi-scale, multidimensional issues of sustainability, a variety of theoretical frameworks and methods is required. A fundamental shift is also needed in our attitude to integrate technology to make better decisions and improve the quality of life. VALUE 2020 sought to create an interface between researchers and practitioners in order to exchange new perspectives and ideas which are essential to improve the quality of the urban environment. It provided a forum where participants drew intelligent insights through group discussions with leading professionals and researchers. This conference brought together several theories and methodologies towards assessing the transition from the present facts to future possibilities or from 'where we are' to 'where we are going' in terms of social, spatial, economic and environmental qualities of urban development.

February 2020

N. Devi Prasad
Conference Chair, VALUE 2020
VIT, Vellore, India

Contents

About the Editors

Dr. K. Thirumaran is an Architect from India. Since 1990, he has been practicing architecture and made many consultancy services for various projects. He is working as Professor in the Department of Architecture, National Institute of Technology, Tiruchirappalli (NITT). He was the Head of the Department (HoD) in Architecture at NITT. He had served as the Member of BOG (Board of Governors) in NITT. He is carrying funded research project from TEQIP. He organized the first National Conference on Recent Trends in Architecture and Civil Engineering towards Energy Efficient and Sustainable Development—2019 (NCACESD 2019) in the Department of Architecture, at NITT. He has organized many national-level workshops for the architecture students in building construction techniques and software skill developments. He authored the book "Impediments of French Architecture Facades in Contemporary Buildings of Puducherry" in the LAP LAMBERT Academic Publisher, UK. His research interest extends in the areas of Built Heritage and Urban Conservation of various historical temple towns and heritage towns, and also the sustainable built environment. Currently, many research scholars are doing their Ph.D. under his guidance. He has published more than 35 research papers in the peer-reviewed journals. He is also a reviewer in various international journals.

Dr. G. Balaji obtained an undergraduate degree in Bachelor of Architecture from the Thiagarjar college of Engineering, Madurai, Tamil Nadu, affiliated to Madurai Kamaraj University, Tamil Nadu, India. He completed his Masters in Architectural Conservation from School of Planning and Architecture, New Delhi. He completed his Ph.D. from the Department of Architecture, National Institute of Technology, Tiruchirappalli, Tamilnadu, India. His research interests include place attachment, morphological changes of channels and physiochemical variations in river water. He was a project coordinator for the joint venture with Columbia University on water urbanism, Madurai, in the year 2015 and was also an invited expert as a jury member for Urban design Project at Columbia University, New York. His contribution as a project director towards the development of the Madurai city includes preparation of the Detailed Heritage Project report for the Madurai city under Jawaharlal Nehru National Urban Mission Renewal in the year 2008 and SMART city mission in the year 2016. He was a technical advisory member for the restoration of the Thousand Pillar Hall in Meenakshi Amman temple in the year 2007 and currently a regional screening committee member for restoration works of Heritage temples across four districts in the state of Tamil Nadu. His recent publication includes "Evaluating the impact of Impervious surface cover of the Madurai Urban Sprawl on River Vaigai through physio chemical parameters," *Environment and Urbanization ASIA*, and "Investigating the impact of Urban setting on Girudhumal River in Madurai," *Journal of Environmental Engineering and Science.*

Prof. N. Devi Prasad is a registered architect and urban designer and has practiced as a professional for over thirty years. He holds a Master's Degree in Architecture in Urban Design from the School of Planning and Architecture, New Delhi. He has been in the profession of architecture and planning for over three decades. His works have extensively focused on architectural projects of different scales, including residential developments, institutional campuses, business workspaces, laboratories and large scale industries including those in Special Economic Zones. As an urban designer registered with the Institute of Urban Designers, India, he has substantial experience handling issues confronting urban scenarios. He was involved in several research studies associated with Varanasi Ghats in Uttar Pradesh and historic precincts of Fort Cochin and Mattancherry in Kerala. In addition, he also worked on projects related to the temple town of Kancheepuram, the historic religious core of Mylapore in Chennai and the World Heritage site of Mamallapuram in Tamil Nadu. He authored a monograph on the historic area of Fort Cochin and an operational document for the Government Municipal Administration Department for the provision of public sanitation services in Tamil Nadu. As an academician, he has been a guest lecturer and a thesis review critic for over twenty years at several architecture schools in Bengaluru, Chennai and New Delhi. He also conducted several faculty development programmes focusing on professionalism in teaching and effective learning methods. He is a strong proponent of Applied Immersive Learning and Collaborative Practices between governing bodies and academic institutions in achieving urban development goals. He is an Executive Committee member of the Chennai Centre of the Indian Institute of Architects and an Executive Member of the National Council of the Institute of Urban Designers, India. As a Professor and Director of the School of Architecture, he has headed the administrative and academic activities of the school from its inception in 2015 and is a member of the Academic Council of VIT, an Institution of Eminence recognised by the Ministry of Human Resources Development, Govt. of India.

Urban Landscape

Peri-Fusion: A Design Strategy for Integrating Densified Housing and Agriculture Within Peri-Urban Zones

Sarah Morris and Hans-Christian Wilhelm

Abstract Low-density, suburban dwelling models are characterising extended areas on the fringes of New Zealand (NZ) towns and cities. The resulting (sub) urban sprawl has been proven unsustainable due to high resource consumption and increasing housing cost. Surveys of housing preferences indicate the continued desire for the standalone home. Increased land usage for settlement conflicts with other land use, including agriculture and green spaces. A cornerstone of NZ economy and agriculture influences housing aspirations, e.g. lifestyle farms. This conflict can also be observed in other countries, where housing infrastructure (for water, waste, food) is insufficient or unsustainable. For peri-urban zones, strategies for low-rise, high-density housing are well established, but lack implementation. Strategies for reducing greenfield developments have been explored, mostly in terms of policy-making. And, concepts for circular housing infrastructure are currently being tested. The lack of uptake can be addressed with a perspective centred on housing design. This research establishes design strategies for integrating densified housing, associated food/waste cycles and preservation of green spaces by way of using agricultural features and motifs, in response to NZ housing preferences. Typological research is combined with research through design, using spatial features derived from agriculture to enhance sustainable housing. The design strategies were tested on a case study site in Blenheim, NZ. The novel design resulted in 38.9% more dwelling units, 34.8% more green space, and a circular water and nutrient model, when compared to an identical adjacent site. This demonstrates the potential of cross-programming/interweaving housing and agricultural landscaping in response to environmental resilience as well as societal housing aspirations.

Keywords Housing · Circularity · Urban agriculture · Landscaping

S. Morris (✉) · H.-C. Wilhelm
Victoria University of Wellington, Wellington, New Zealand
e-mail: sarahapp13@hotmail.com

H.-C. Wilhelm
e-mail: hans-christian.wilhelm@vuw.ac.nz

K. Thirumaran et al. (eds.), *Sustainable Urban Architecture*,
Lecture Notes in Civil Engineering 114,
https://doi.org/10.1007/978-981-15-9585-1_1

1 Problem Statement and Relevance

Low-density housing and suburban models of settlement are characterising extended areas outside central business districts (CBDs) of New Zealand (NZ) towns and cities. As in other countries with a cultural preference for the single family house (including Australia, the USA and parts of Europe), the resulting sprawl has been proven unsustainable due to high resource consumption in terms of land, infrastructure and energy, as well as contributing to high housing cost (Abrahamse et al. [1]. Nonetheless, surveys of NZ housing aspirations indicate the desire for the stand-alone home, including private garage and backyard, as a preferred dwelling model and lifestyle [6]. This relates back to the nineteenth century, when settlement promoters advertised New Zealand with a rural vision—an image supported by NZ Government well into the twentieth century [10]. Accordingly, ample space (interior and exterior) for the individual family home, privacy and access to green spaces are the perceived attributes of this rural inspired vision of housing.

Despite a low average population density, increasing land usage for housing and settlement conflicts with land available for green spaces and agriculture. This has been identified a key issue in the Environment Aotearoa 2019 Summary Report, which states that urban land increased by 10% between 1996 and 2012, and "between 1990 and 2008, 29% of new urban areas were on 'versatile' land" [18].

Agriculture is not only a cornerstone of NZ economy, but also embedded in NZ social history [22]. In fact, lifestyle farming (i.e. a single family house on a plot larger than required for housing alone and used for extensive farming for leisure and lifestyle) can be observed in many NZ non-urban areas. In 2011, 873,000 ha was used for 175,000 lifestyle blocks [18]. Even though this land is legally reversible back to agricultural land, the "increased value of the property per hectare", caused by construction of large detached homes [2], makes this an economically unattractive and hence rare option. An urban phenomenon remotely connected to this reversing of land, community gardens, community gardens for leisure as well as contributing to food supply are gaining popularity in NZ as an element of a lifestyle towards individual sustainability [23].

Areas characterised by sprawl are often located on the outskirts of towns and cities, where this transitional area from urban to rural contexts can be defined as the peri-urban zone [12], highlighting its circumferential or peripheral character. For areas outside CBDs, scholarly and professional researches have successfully identified the benefits of and strategies for low-rise, high-density housing, or medium-density housing (MDH) [5]. However, societal reservations against this dwelling model contribute to a lack of implementation [6]. Similarly, strategies and policies for reducing greenfield depletion caused by building developments have been explored, but often have limited impact. In terms of housing infrastructure (for freshwater and wastewater, as well as nutrients and food), concepts for circular models have been investigated and are being tested at small scale, for example, Roof Water-Farm in Berlin [17]. This is of particular interest to NZ towns and peri-urban zones, as often there are no centralised sewage systems, but septic tanks used instead. This research aims to address and turn into opportunities, the scarcity of housing infrastructure, lack of uptake of densified housing models, and reconciliation with NZ housing aspirations. For this purpose, existing concepts developed in isolation need to be integrated through the lens of housing design.

In terms of housing aspirations founded on nineteenth-century NZ settlement models, the perception of privacy and autonomy is associated with owning and inhabiting a detached home, with unobstructed views of green spaces. However, research is lacking on how this perception can be achieved with less land usage and preserving green spaces and/or agricultural land. This is of particular relevance in peri-urban zones.

2 Objective

This paper aims to establish design strategies and test solutions on three different design scales for integrating densified housing, associated food/waste cycles and preservation of green spaces/agricultural land within peri-urban zones. The use of agricultural features, motifs and materials not only can respond to NZ housing preferences, but can enhance the quality of densified housing by redefining and blurring the boundaries between interior and exterior green space, and establishing degrees of privacy between dwellings. Additionally, it is expected that in this way housing can be embedded within a natural resource cycle (water, waste, nutrients) and thus becoming more resilient.

Proposition

This paper proposes that spatial features and motifs derived from agriculture and traditional landscape design can be used and applied to enhance quality of densified housing. This includes a well-defined transition between public, semi-public and private areas, whereas vice versa housing under this model contributes to increasing biodiversity and limiting loss of versatile, unsealed land. This strategy allows for a cross-programming, whereby housing and agricultural landscaping are interwoven to form new development models in response to environmental resilience as well as societal housing aspirations. Fusing these elements in peri-urban zones is captured in the novel term “peri-fusion”.

3 Methodology

Based on a literature review, typological research is combined with research through design, using spatial features derived from agriculture to enhance sustainable housing. In a first methodological step, the relevant literature and case study projects are reviewed for common types of low-rise, high-density housing, integration of green spaces, and agricultural landscape design strategies.

Subsequently, the design strategies were tested on three incremental design scales (master plan, street scale, house scale) for a case study site in Blenheim, NZ. The resulting master plan design has been measured against a set of parameters, indicating land use/density and relation sealed versus unsealed/green land. The master plan site was then compared to an identical adjacent site with existing suburban housing. The findings demonstrate the potential of cross-programming/ interweaving housing and agricultural landscaping to form new development models in response to environmental resilience as well as societal housing aspirations.

4 Literature Review and Typological Research

NZ peri-urban zones are lowly populated and characterised by one- to two-storey buildings in repetitive or loose arrays. Therefore, low-rise, high-density housing has been the starting point for the literature review. Due to the broad scope of this subject, the review focuses on options for integrating densified housing and surrounding landscape. Other focus areas include strategies for integrating urban housing and agricultural space, and design strategies for decreasing sealed land use. Agricultural landscape design features are reviewed through a typological lens, rather than for agricultural productivity. The reviews aim to establish a phenomenology of spatial interventions associated with agriculture, for the purposes of sustainable and desirable housing—a strategy for fusing agricultural/green and residential spaces. This includes the opportunity of achieving qualities of rural or non-urban living, as desired by many New Zealanders.

4.1 Low-Rise, High-Density Housing

European Housing Concepts discuss a variety of fusion concepts between housing and the landscape [11]. A precedent which demonstrates housing qualities aspirational for NZ is the Looren Housing Complex, by Metron AG architects. This is a series of semi-detached and row houses, where dividing walls extend into fence lines, creating private backyards with uninterrupted views into common green areas. Locating the car park partially underneath the row houses decreases the amount of sealed area dedicated to private vehicles. Subtle-level changes and a hierarchy between dwellings and semi-public spaces create an effective delineation between public and private spaces.

"High-density, Low-rise—A Challenge for Dwelling Landscapes in the Netherlands" proposes a set of design tactics for integrating landscaping with housing at urban fringes [13]. These strategies, derived from successful case studies, are explored from urban planning to architectural design scales.

For the purposes of this paper, a housing programme has been established on the basis of statistical population forecasts. Statistical data also reveals that average NZ per capita internal gross floor area is currently among the largest in the world [8]. In response to this, the housing design strategy is based upon three compact low-rise housing typologies, i.e. detached, semi-detached and row housing, with the potential to achieve higher densities than current NZ practice. Floor layouts have been based on recommendations for minimum internal dwelling areas from Auckland Design guides, resulting in much more compact yet functional layouts [3]. The features for privacy/boundaries identified in the precedents then have informed the detailed housing designs for the Blenheim case study design.

4.2 Low-Rise, High-Density Housing

Dewey Thorbeck's *Architecture and Agriculture: A Rural Design Guide* provides a general framework for agricultural and urban integration within peri-urban zones. He expresses the importance of design when integrating urban and agricultural contexts, stating that "rural design is a new design discipline that tries to transcend the dichotomy between urban and rural to integrate and make connections at both the macro and micro" [21].

Complexity in Urban Agriculture: The Role of Landscape Typologies in Promoting Urban Agriculture defines urban agricultural, spatial and programmatic typologies. These include community gardens and community managed farms, demonstration gardens, animal husbandry, aquaponics, greenhouses, resource centres, food processing and retail [20]. They form potential agricultural spaces within urban environments, with the opportunity to produce food and recycle waste.

"Village Homes" is a housing development by Michael (Mike) and Judy Corbett, which demonstrates different levels of integration between dwelling and agriculture [9]. The master plan is considered successful due to innovated planning of central green spaces, agricultural fringes and corridors, nonlinear pathways, sunken communal car parking, semi-public to public zones. Level change between dwellings to improve privacy is also identified as effective design features.

A Dictionary of Agriculture and Land Management and *A Dictionary of Landscape Architecture* proposes and defines landscaping design features within or derived from agricultural contexts. The typologies include a ditch with a gradient side and a retaining wall side, a feature known as ha-ha wall, banks, sunken fences, fences, dry stone wall, hedgerows, ditches, bridleways, green lanes and field margins [15, 19].

For the purpose of understanding contemporary rural design in NZ, various modern agricultural builds have been investigated to establish commonly used agricultural materials and motifs in NZ, including timber, concrete blockwork and corrugated iron. These materials are used equally for utilitarian buildings as well as residential signature architecture, sometimes forming an agricultural landscape, where material choices are often driven by low cost, easy availability, low maintenance and functionality [4].

4.3 Literature Review Summary and Design Strategies

The literature reviews on low-rise, high-density housing and landscaping have shown design tactics suitable for the research objectives, including uncompromised views into green space, while providing boundaries through level changes. Similarly, establishing hierarchies between dwellings and external spaces can achieve privacy. Both views and privacy are considered qualities for aspirational housing for New Zealanders. While the focus of this paper is not on food production, the literature on urban agricultural integration indicates the types and dedications of agricultural spaces. Typological review of landscaping features related to agriculture shows that these elements can contribute to fuse landscaping with housing. These spatial features have been adopted as the basis of the peri-fusion design strategies, i.e. how dwellings can integrate with agricultural green space within peri-urban zones.

Design Strategies

A selection of relevant design objectives and corresponding landscape design features has been synthesised and translated into peri-fusion design strategies (see Fig. 1). While they are derived from agricultural contexts, here they are used for the purpose of the envisaged integrated housing and agriculture design. Strategies include landscape interventions (such as level changes, ha-ha walls, banks, sunken fences, hedgerows, ditches), shared access ways (bridleways, green lanes and field margins) and rural motifs/materials. The landscaping interventions in particular are used to control degrees of privacy, creating soft boundaries, maximising the extent and perception of green space, and responding to rural housing preferences, including private and community gardens (Table 1).

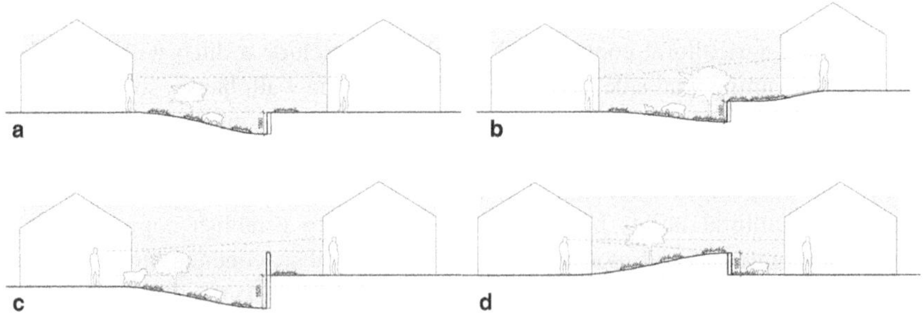

Fig. 1 **a** Ha-ha wall, **b** level change ha-ha wall, **c** tall ha-ha wall, **d** reversed ha-ha wall

Table 1 Agricultural design strategies

Objectives	Design strategies	Diagrams
Degrees of privacy, uninterrupted views	Agricultural boundaries (ha-ha wall), level change/ rolling landscape	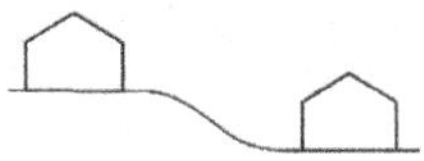
Maximising green space, natural resource cycle	Ha-ha wall (water collection), extensive grazing creating nutrients for intensive horticulture	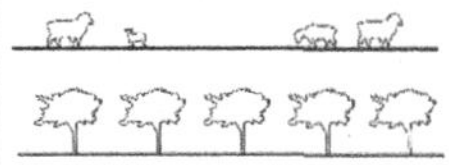 
Rural vision, rural desirability	Agricultural motifs and materiality, agricultural boundaries	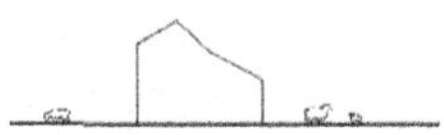
Agricultural integration	Shared access ways, agricultural boundaries	

In this context, the so-called ha-ha wall was identified as highly versatile as it has different properties on either side. This agricultural spatial boundary emerged in the eighteenth century in Britain as a physical barrier for livestock, while leaving lines of sight within the landscaped garden uncompromised [15]. As a partly invisible and adaptable barrier, this element has been used as a generator for privacy and human access control, as well as in its original role as a barrier to extensive livestock (see Fig. 1a–d). In the envisaged peri-fusion housing model, in fact the ha-ha wall can be used in both ways, since sheep could be used for both the purpose of maintaining the grassland and contributing to a circular waste/nutrient model in the form of creating compost. Research indicates that agricultural models combining crops/horticulture and some limited livestock have the highest sustainability benefits [7]. However, in this context, both horticulture and livestock are used in extensive forms, contributing to a broader sustainable way of operating and maintaining housing, rather than to food production (alone). Additionally, a system of ha-ha walls with ditches could be used for both water collection and, in combination with reed or other suitable planting, grey water purification. Figure 1 shows the adaptive iterations of the ha-ha wall, where the agricultural boundary exists not just between dwelling and livestock, but also between houses. Heights, ground-level change and sloping were tested to alter the levels of privacy and sightlines between residents, while controlling livestock to one side.

5 Blenheim Case Study

In a second methodological step, the design strategies for dwellings and landscaping were fused and applied to a case study site in Blenheim, NZ. Blenheim is a growing town with a population of 31,600, currently experiencing substantial urban sprawling in its peri-urban zones. It is also suitable due to the established contribution of horticulture and viticulture to the local economy and culture. Alongside small livestock, this makes up the majority of Blenheim's rural land use, including the peri-urban zones. The presence of a horse race course, recently fallen into disuse, has been used as an additional opportunity. Rather than a greenfield site, former Waterlea Racecourse has been chosen as the site for the case study design (see Fig. 3). In addition to developing and refining the design strategies for the purpose integrating densified housing and agriculture, the success of the strategies has been quantitatively assessed by way of comparison to an equally sized portion of existing suburban housing. For this purpose, quantitative parameters were established, including total number of dwellings, dwellings per hectare, average gross floor area per dwelling, sealed versus unsealed surface area. These parameters have been examined for two flat, immediately adjacent sites of each 31.14 ha (or 311,400 m^2), that is, the former Waterlea Racecourse (plot B) and a residential area (plot A) adjacent to the west. Additionally, by way of design research, the suitability of the peri-fusion design strategies mentioned previously has been assessed qualitatively against the objectives mentioned in Sect. 2. This includes iterations of the ha-ha wall, level changes or other forms of soft or invisible boundaries, and pathways shared between human, vehicles and animals (see Fig. 2).

Prior to the quantitative comparison, design research has been undertaken at different design scales, including developing design solutions at master plan scale (1:2000), street scale (1:1000) and house scale (1:100). Subsequently, the quantitative parameters of sealed versus unsealed land have been examined, using a

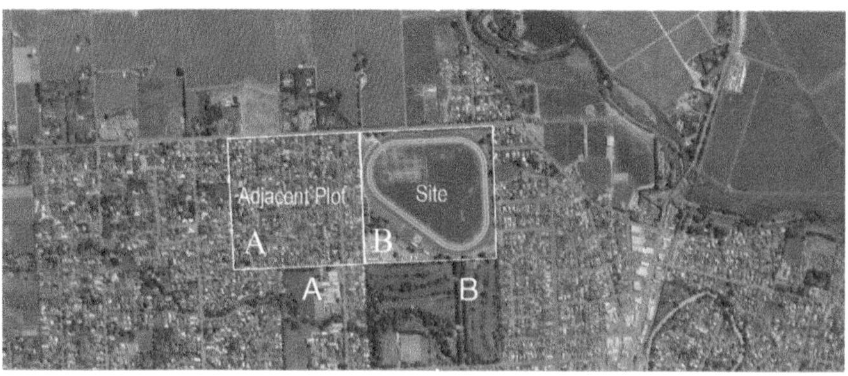

Fig. 2 Aerial image of northern Blenheim, highlighting Waterlea Racecourse (site, plot B) and reference area (plot A). Based on aerial picture by Marlborough District Council. https://maps.marlborough.govt.nz/smaps/?map=c6bbc97937834f46a2a3a18d76e79db8

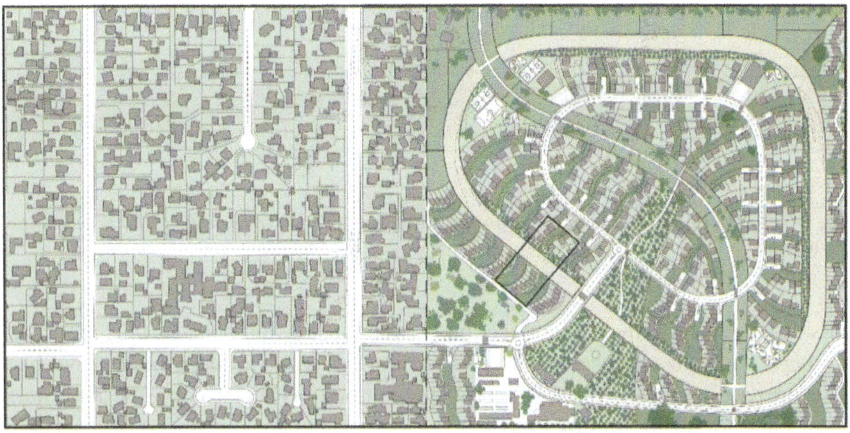

Fig. 3 Reference area (plot A, left) and master plan of case study peri-fusion housing model (plot B, right), not to scale. Extent of street scale plan (Fig. 4) indicated in black outline. Based on data by Land Information NZ. https://data.linz.govt.nz/layer/101290-nz-building-outlines/

finalised master plan. Figure 3 illustrates the measured areas for the sealed (white = hard surfacing and vehicle access, and grey = all building footprints) versus unsealed surfaces (beige = soft surfacing and green = soft landscaping, vegetation). The design strategies as applied to the final housing model have created densified clusters of housing, while maintaining views into preserved rural landscape and defining public, semi-public and private spaces through level changes (see Fig. 4) and other soft boundaries. For the houses, parallel lateral walls have been used to generate compact floor plans. Extending these walls from inside to

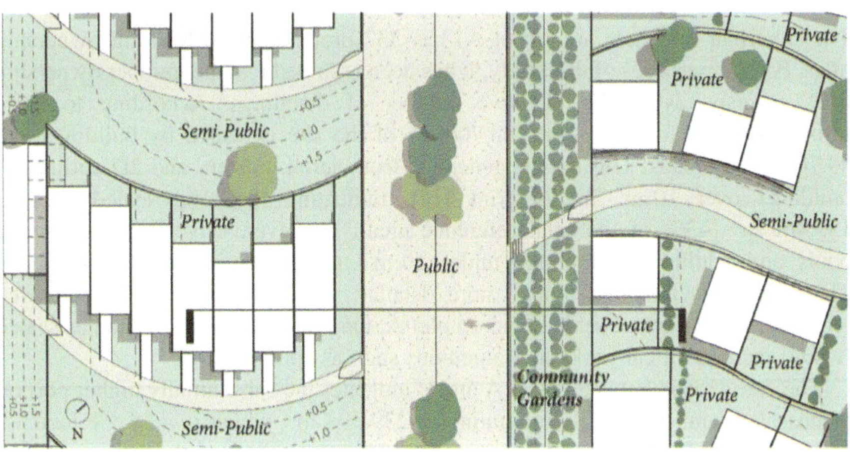

Fig. 4 Street scale, plan, indicating row housing, semi-detached and detached houses. Walls part of ha-ha walls or houses indicated as black lines. Extent of section (Fig. 5) indicated, not to scale

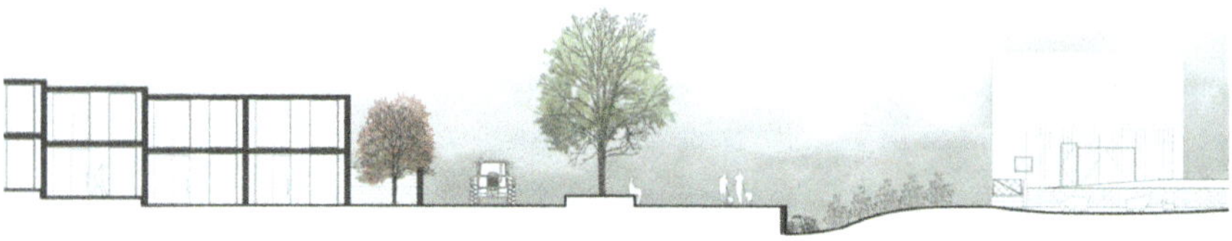

Fig. 5 Street scale, section, indicating former horse race track, row housing and detached houses, not to scale

outside has enabled a visual connection between inside and outside, thus increasing the perception of ampleness and space within houses. The floor plans have been developed to allow for three different orientations (as dictated by the racecourse figure) and tested for sun exposure.

Master plan

The developed master plan comprises 468 dwelling units in row houses (199 three bedroom units), semi-detached (105 one and 105 two bedroom units) and detached houses (69 four bedroom units). In addition, existing buildings have been conceptually devised to include a range of public and agricultural amenities. Large amounts of space are dedicated to agricultural and horticultural practices (see Fig. 3).

6 Findings

Quantitative Findings

The suburban reference area (plot A) has 337 property plots/dwelling ownership titles. For the purposes of this study, it has been assumed that the number ownership titles corresponds to the effective number of dwellings. According to Land Information NZ footprint data, 83,706 m^2 of the site is sealed by buildings and 49,763 m^2 by roads. This data is generated from aerial imagery and 2D outlines of buildings over 10 m^2 in footprint [14], indicating the total sealed area as 133,469 m^2 (43% of the total reference area). However, driveways, swimming pools and additional structures under 10 m^2 are not considered in this data. Therefore, it is assumed that a more detailed survey would result in an even increased percentage of sealed land on plot A. On plot B, areas calculated as sealed (white) include all car parks and amenities such as sports courts.

The comparison between plots A and B indicates the substantially higher portion of unsealed land on plot B, amounting to 239,394 m^2, which is a 34.8% increase when compared to 177,631 m^2 on plot A. This demonstrates the success of the

Table 2 Comparative study summary table

	Existing adjacent (plot A)	New case study (plot B)
Number of dwellings	337	468
Dwellings per hectare	10.8	15
Average gross floor area/ (dwelling)	175 m^2	80.5 m^2
Total sealed surface area	133,469 m^2 (43%) of 311, 400 m^2	71,965 m^2 (23%) of 311, 400 m^2
Total unsealed surface area	177,631 m^2 (57%) of 311, 400 m^2	239,394 m^2 (77%) of 311, 400 m^2

developed fusion design strategies. The dirt racetrack in the new housing model (plot B) has been left as an unsealed walkway and bridleway. The number of dwelling units compares 337 for plot A versus 468 on plot B, i.e. an additional 131 units or 38.9% more (see Table 2). The average gross internal floor area per dwelling unit (one dwelling unit per house) on plot A is 175 m^2 (as indicated on Marlborough District Council property files), whereas the proposed design on plot B only shows 80.2 m^2 per dwelling unit, allowing for more unsealed land and an increase of dwellings per hectare [16]. Although less than half the size, these sizes not only are in accordance with the pertinent NZ housing guides [3], but also address the fact that the current average per capita internal dwelling area in NZ is among the largest in the world [8].

Implications

Peri-fusion as a set of design strategies is not limited to New Zealand. The notion of the free-standing family house in a green landscape setting as a privileged, desirable housing model seems to spread beyond the countries initially mentioned—alongside with the suburban sprawl it induces. Versatile agricultural land is globally under pressure, and the significance of circular economic models in response to shortening natural resources has been broadly recognised. Therefore, the underlying concept of integrating housing and landscaping with agricultural elements, of embedding housing in a natural resource cycle, is relevant on a global scale. The design strategies and tactics developed could be adapted to suit varying cultural and physical environments. In terms of housing density, what is regarded high density in New Zealand might be considered low density in another location. Further research in this field should include interweaving architecture and agricultural landscaping at a more detailed scale, considering not only space design, but details of water, nutrient and materials cycles. Here, peri-fusion and its design strategies can be a robust and versatile framework.

References

1. Abrahamse W, Stuart K, Witten K (2011) Growth misconduct? Avoiding sprawl and improving urban intensification in New Zealand. In: Witten K, Abrahamse W, Stuart K (ed) Steele Roberts, Auckland, N.Z
2. Andrew R, Dymond J (2013) Expansion of lifestyle blocks and urban areas onto high-class land: an update for planning and policy. J Roy Soc New Zealand 43(3):128–140. https://doi.org/10.1080/03036758.2012.736392
3. Auckland Council (n.d.) Apartment space—auckland design manual. Retrieved 10 December 2019, from Auckland Design Manual website: http://www.aucklanddesignmanual.co.nz/sites-and-buildings/apartments/guidance/the-building/apartment-layout/apartment-space
4. Barnbuilders (n.d.) Farm buildings and sheds. Retrieved 14 October 2019, from Barnbuilders Tim Magill website: https://www.barnbuilders.co.nz/FARM+BUILDINGSSHEDS.html
5. Bryson K (2017) Defining medium-density housing. BRANZ, Porirua, New Zealand
6. Bryson K (2017) The New Zealand housing preferences survey: attitudes towards medium-density housing. BRANZ Ltd., Judgeford, New Zealand, p 45
7. Coffey L (2014) Integrating livestock and crops: improving soil, solving problems, increasing income, pp 1–16. NCAT
8. CommSec (2018) Australian home size hits 20-year low: CommSec home size trend report. Australia
9. Dingemans DJ (2018) The village homes subdivision in davis: origins and evolution of "a better place to live". Yearbook Assoc Pacific Coast Geogr 80:13–40
10. Ferguson G (1994) Building the New Zealand dream. Dunmore Press with the assistance of the Historical Branch, Department of Internal Affairs, Palmerston North, N.Z
11. Gelsomino L, Marioni O (2009) European housing concepts 1990–2010. Editrice Compositori, Bologna
12. Iaquinta D, Drescher AW (2000) Defining the peri-urban: rural-urban linkages and institutional connections. Land Reform Land Settl Cooperatives 8–27
13. Kuitenbrouwer PAM, De Saeger R (2015) High-density, low-rise—a challenge for dwelling landscapes in the Netherlands: architectural research by design as a process towards incorporated typologies. Boundaries|Encounters|Connections: papers presented at the housing and welfare conference, Copenhagen, Denmark, 7–9 May 2015, 7–9 may 2015
14. Land Information New Zealand (2019) NZ buildings data dictionary. Retrieved 23 November 2019, from Land Information New Zealand website: https://nz-buildings.readthedocs.io/en/latest/introduction.html
15. Manley W, Foot K, Davis A (2019) Dictionary Agric Land Manag. https://doi.org/10.1093/acref/9780199654406.001.0001
16. Marlborough District Council (2019) Property files online—Marlborough district council. Retrieved 10 December 2019, website: https://www.marlborough.govt.nz/services/property-files-online
17. Million A, Bürgow G, Steglich A (2018) Roof water-farm. Technische Universität Berlin, Berlin
18. Ministry for the Environment & Stats NZ (2019) New Zealand's environmental reporting series: environment Aotearoa 2019. Retrieved on 09 December 2019 from https://www.mfe.govt.nz/environment-aotearoa-2019-summary
19. Morrow BH (1987) A dictionary of landscape architecture, 1st edn. University of New Mexico Press. Mexico, Albuquerque
20. Napawan NC (2016) Complexity in urban agriculture: the role of landscape typologies in promoting urban agriculture's growth. J Urbanism Int Res Placemaking Urban Sustain 9 (1):19–38. https://doi.org/10.1080/17549175.2014.950317

21. Thorbeck D (2017) Architecture and agriculture: a rural design guide/ Dewey Thorbeck. Routledge, Milton Park, Abingdon, Oxon
22. Thornton GG (1986) The New Zealand heritage of farm buildings. Reed Methuen, Auckland N.Z
23. Webb V (2017) Good gardeners and bad plants: governing health in community gardens. New Zealand Sociol 32(2):117

Conservation and Protection of Peri-Urban Rural Landscapes from the Impacts of Urbanization: Case Study of Manimangalam, Mahanyam and Malaipattu Villages in Manimangalam Watershed

Kumareswari Rajendran

Abstract Rural farmland preservation, open spaces, water bodies and woodland in the suburbs strongly contrast the borders of urban and peri-urban areas. High loss of agricultural lands and land plots as dumping grounds found in suburban due to unplanned urban development, which in turn would impact economic growth and contribute to poor quality of life in the suburbs. The suburban fringes have a complex political and demographic character. Industries and development of real estate are transforming rural land as it is available at affordable prices, agricultural land is being transformed into barren land and common ground, and open spaces are being gradually invaded. Manimangalam watershed region in the sriperambadur taluk in the suburban area of Chennai is one of those places that faces many challenges due to urban sprawl and industrial growth, Manimangalam watershed with a chain of 56 lakes, Mahanyam reserve forest and Malaipattu hills have a high quality of scenery in it. The change in land use is due primarily to the expansion of the corridors of urbanization, industrialization and transportation. Green and brown fields split the high scenic quality of rural landscapes and they need to be given great attention. Cultural, ecological, economic and agricultural values and the ideals of leisure must be maintained and the conservation and growth potential must be evaluated and included in suburban development urban planning. Urban growth boundary and smart growth of peri-urban areas and the conservation of rural landscapes will encourage sustainable urban development in the future.

Keywords Conservation · Urbanization · Peri-urban · Rural landscape and urban growth boundary

K. Rajendran (✉)
Faculty of Architecture, Dr. M.G.R. Educational and Research Institute, Chennai, Tamilnadu 600095, India
e-mail: ar.kumareswari@gmail.com

K. Thirumaran et al. (eds.), *Sustainable Urban Architecture*,
Lecture Notes in Civil Engineering 114,
https://doi.org/10.1007/978-981-15-9585-1_2

1 Introduction—Urbanization and Its Impacts and Significance of Rural Landscape

Urban areas that have expanded physically at a rapid rate in recent decades as a result of developments in transportation and industrialization, massive population growth and increased migratory flows from rural areas to urban areas are increasing day by day.

Similarly, urban development pressure lies on the peri-urban and rural fringes.

Rural landscapes with lots of water bodies, forests and rich biodiversity with flora and fauna abound in natural resources.

Urbanization impacts seen in rural peri-urban landscapes include disturbances to the ecosystem, degradation of water bodies and exploitation of natural resources.

Rural ecosystems in peri-urban areas provide a wide range of ecosystem services, such as provisional, regulatory, cultural and habitat services.

Agricultural lands in rural areas serve as the productive landscapes and food providers for the urban areas to meet surplus demand. Water scarcity during summer is balanced by the supply from the water bodies of these peri-urban regions and also they act as the buffer zone for flood hazards during heavy rains through their wetlands and vegetation and reduce the run-off. Vegetation and forest in the rural areas act as a breathing and buffer zone against urban and industrial pollution. These offer recreation and nature trail opportunities and long-distance driving opportunities in green areas as an escape from urban traffic. They play a major role in providing a balance ecology by provision of habitat for various flora and fauna and thus enhancing the biodiversity in the region.

Urbanization is a major problem in recent decades, as economic growth and development have concentrated in cities; the population density in urban areas is multiplying worldwide [1]. Urban areas that are unable to cope with the capacity to hold large populations of migration day by day will find a solution by extending urban boundaries to suburban areas that affect rural landscapes in urban areas, and this exorbitant urban spread will result in the deterioration of rural landscapes in urban areas, and if there is no managed or urban growth boundary, the loss of rural areas will be limited.

1.1 Significant Services Provided in the Rural and Peri-Urban Areas by Open Spaces and Agricultural Land

Open areas and agricultural lands in peri-urban and rural areas serve as the natural filtration area (ponds, wetlands and lakes), water storage and run-off (wells, lakes and rain water harvesting structures like farm ponds), improve the air quality and act as buffer zone against pollution, serve as carbon capture and carbon sequestration zone (forest and wetlands), recreational zone and high scenic value and enhance esthetics of the region (Fig. 1).

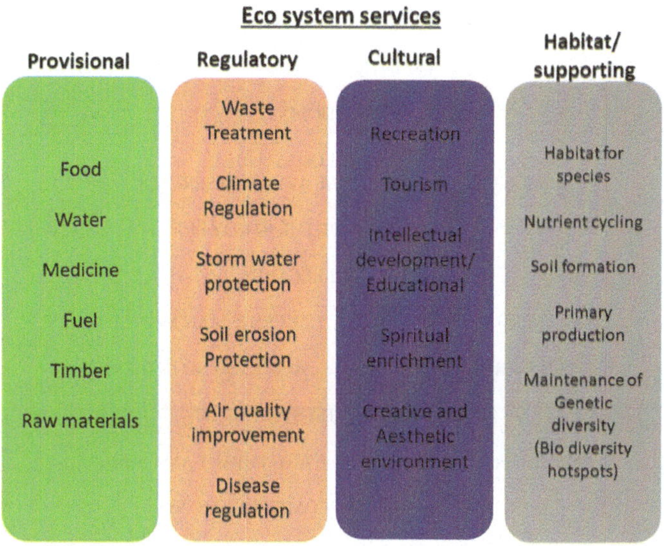

Fig. 1 Ecosystem services provided by the rural landscapes

2 Research Study Approach

See Fig. 2.

3 Methodology

1. Data collection
   Primary data collection—observation and field visit, site reconnaissance and field survey (rural people). Literature study—newspaper articles, research papers are used as reference for further understanding the problem as well as the study area (history, natural layer and man-made layers' study).
   Secondary data collection—maps and data collection from government. Departments—PWD, agricultural departments, TWAD board, forest department, information from officials and records, NGO data and NRSC and BHUVAN and Google earth satellite maps are used for study data.
2. Data analysis and inferences
   Current land use maps and old maps to be over layed and land use changes Analyzed following Ian Mc Harg Overlay method and analyzed [2].

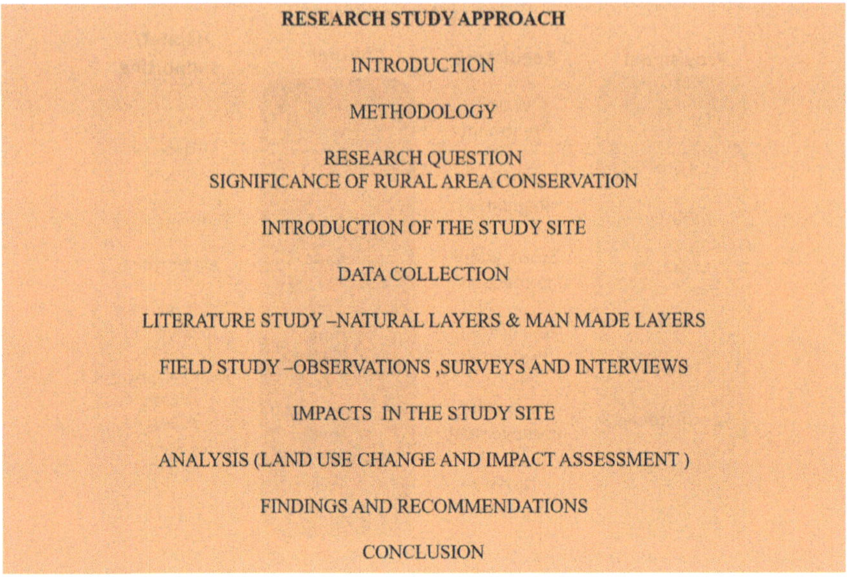

Fig. 2 Research study approach pattern

4 Research Question

- Need to protect the rural and peri-urban landscapes?
- What are the various impacts of urbanization which degrade the rural/peri-urban ecosystems?
- Why rural landscapes ecosystem is significant?
- Why Manimangalam, Mahanyam and Malaipattuis taken as study site?
- What are the impacts found in the site which has turned as the disturbance for the rural ecosystem?
- What are the possible measures to protect the rural landscapes for the future? (Figs. 3, 4)

5 Studysite

Manimangalam, Mahanyam and Malaipattu villages located in the Manimangalam watershed have a lot of landscape value and conservation potential with its natural areas and high scenic value with beautiful hills, forests, sandy lake agricultural areas and a very small population and lush greenery and good quality of space and environment in the urban outskirts of south Chennai.

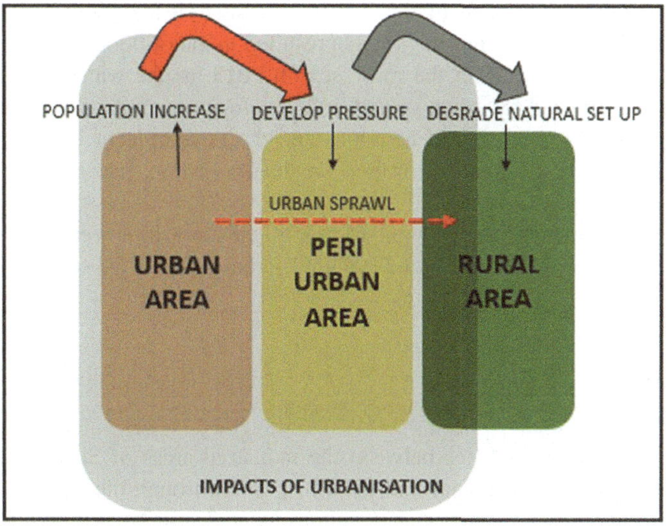

Fig. 3 Need to protect the peri-urban landscapes

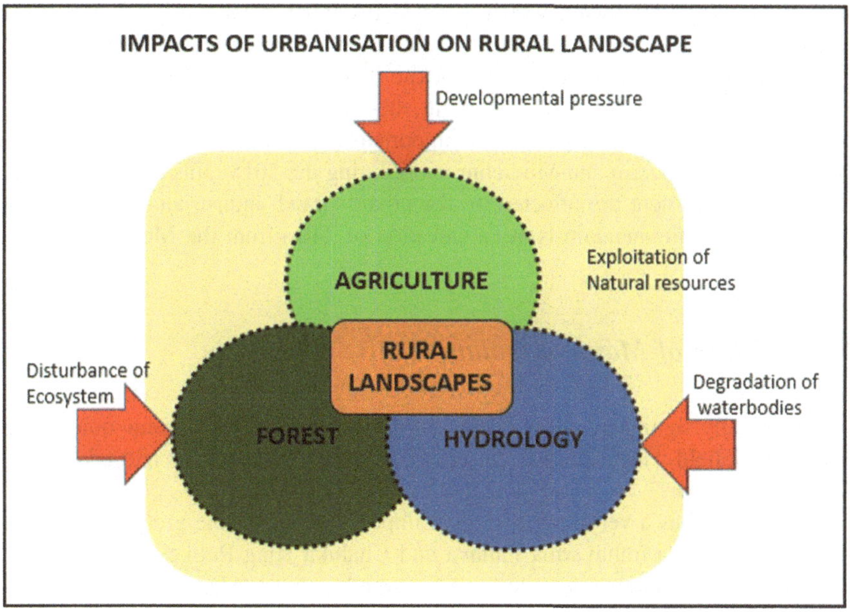

Fig. 4 Various impacts of urbanization which degrade the rural/peri-urban landscapes

Manimangalam watershed of the Adyar watershed at the southern edge of the city of Chennai is an important region that resulted in major floods in the Tambaram and Mudichur regions during the rainy season 2015, as the watershed and catchment areas are affected by urban sprawl, industrialization and land use change. The location of various major urban activities and the development of transport links since the foundation of Chennai in the seventeenth century had a huge impact on the city's modern urban shape.

The urban shape of the Chennai Metropolitan Area has been determined by developments along major roads and rail links radiating from the center of Chennai.

6 Location

Manimangalam watershed lies between the industrial areas of Sriperambadur and the peri-urban areas of Chennai and SH 110 passes through the region.

Manimangalam is a located in Kunnattur city in Tamil Nadu State, India.

Sethupattu, Malaipattu, Karasangal, Mahanyam, Kolathur are the nearby localities to Manimangalam in Manimangalam watershed region.

Manimangalam village is located about 10 km from Tambaram on the western side of Mudichur Highway, which leads to Padappai. Manimangalam watershed region has a tropical climate with high humidity and temperature. Geomorphological formation of Charnockite rocks is found in this region (Fig. 5).

Manimangalam watershed important sub-basin of Adyar watershed at the southern edge of the Chennai city is an important region which resulted in the major floods in the Tambaram and Mudichur areas during the 2015 rains as the watershed region and catchment are affected by the urban sprawl, industrialization and land use change. Manimangalam is at an elevation of 24 m from the Mean Sea Level.

6.1 History of Manimangalam

An historically significant area with rich rural scenery and high scenic quality was once a battlefield where the war between Chalukyas and Pallavas had taken place 1000 years earlier.

This village has a very high historical importance due to the great battle fought here between Narasimhavarma Pallava and Chaluka King Pulikesi II. This battle was fought in Manimangalam during the seventh century AD.

Manimangalam has an ancient temple named Sri Dharmeswarar for Lord Shiva.

Sri Vedhaambigai is the goddess here. According to the inscriptions, this temple is said to have been built by Chozhas. This temple is listed in the inscriptions as Grama Sikhamani Chathurvedhi Mangalam, belonging to Kunrathur Naadu, a division of Puliyur Kottam in the district of Jayamkonda Chozha Mandalam. In

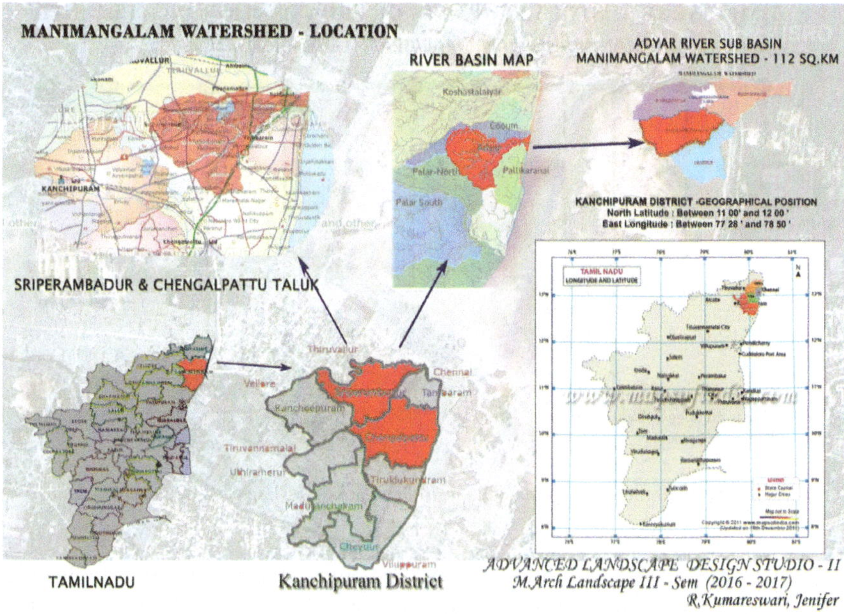

Fig. 5 Location map—Manimangalam watershed region

ancient times, the lands and the temple were donated by the kings to the Vedic scholars to conduct the Vedic rituals for the well-being of the country. These villages are called Chathurvedhi Mangalam (Fig. 6).

6.2 Manimangalam Watershed Region—Manimangalam Village

The Manimangalam watershed consists of 56 lakes of linked chain link reservoirs.

Manimangalam lake, the last large lake in the Manimangalam basin, is a large water body of 3.42 sq.km with a total catchment of 27.63 sq. miles are provided with 10 sluices and 5 weirs with a maximum tank capacity of 225 mcft and due to the impacts of urbanization and developments and encroachments in the catchment and blockage in the drains were the major reasons for flood hazards in remote and urban areas during 2015. Simultaenously during consecutive years of drought after Vardha, Chennai PWD planned to draw water from the massive Manimangalam and Malaipattu lake to meet the drinking water demands of Chennai during summer.

The catchment area of Manimangalam lake is permeated by dwellings and industrial developments have totally disturbed the hydrology of the region. The lake has been broken by the development of SH 110, and the inlet and outlet flow have

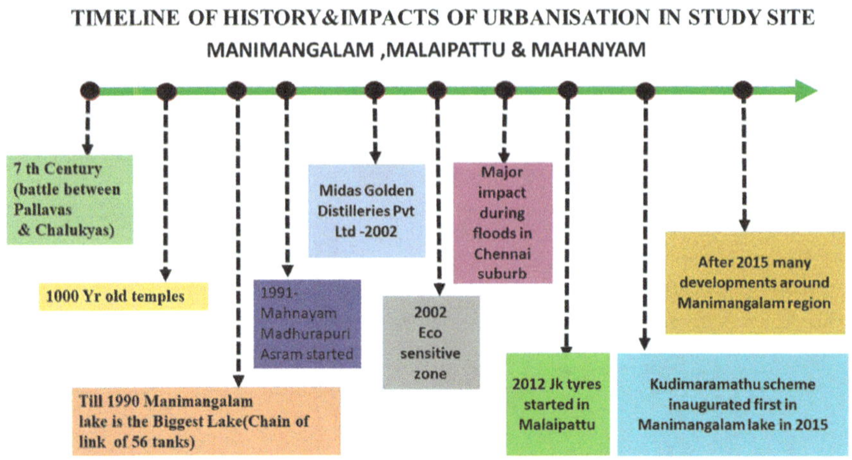

Fig. 6 Timeline of events in Manimangalam, Mahanyam and Malaipattu villages

been interrupted by blockages during road construction and by the development of high-power transmission lines passing through the lake and thus affecting migratory birds arriving at the lake (Fig. 7).

The lake has been an ecosensitive region where a lot of migratory birds come every season as they lie in the Kargili and Vedanthangal bird sanctuary migration routes.

The flood zones of the lake have been impacted by residential developments approval with Neermoozhgum patas, which means that the land will be filled with water during the floods, and therefore, the catchment and hydrology of the area are completely disrupted. Agricultural land has been transformed as fallow land due to the lack of irrigation during the summer season and converted to fallow land, and for the next three years, non-agricultural land has been granted permission for development, with the result that vast agricultural land is converted to fallow land and then made into residential areas due to development in real estate as peri-urban and rural areas.

The area has spectacular cultural landscapes, with 2000-year-old temples and rich vegetation, and very old trees and lakes that have gradually faced a lot of urban problems in recent years.

6.3 *Malaipattu Village*

Malaipattu is a beautiful village with a rich scenic values and good visual quality of hills and 24 m MSL lake with a population of 1250.

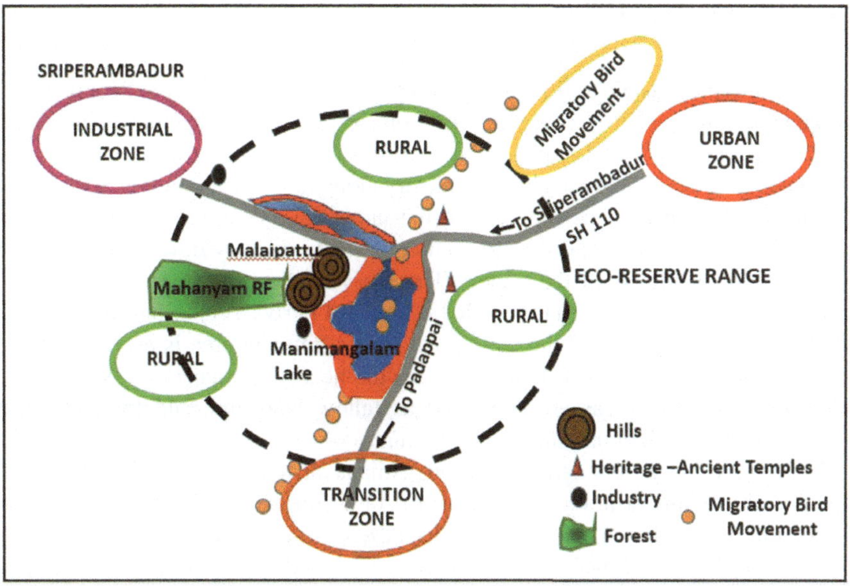

Fig. 7 Schematic sketch of Manimangalam and its neighborhood

Malaipattu lake is one of the lakes from which one of the branches of Adyar river branch starts, and the lake has very sweet drinking water which irrigates agricultural land in the Malaipattu area with beautiful flower gardens and nurseries.

Malaipattu hills are geomorphologically pre-Archean rock source in the hill catchment is deposited in Manimangalam lake. The hill once surrounded by rich tropical dry deciduous trees is now plagued by the erosion of the plates, as a large number of quarries operating illegally eroded a small portion of the hill, resulting in erosion and gravel quarrying from the hills and the excavations resulted in a deep holes and the abandoned quarries become an environmental threat to the peoples, animal and plants of the region in the Pedi plains of Malaipattu hills.

The quarries are extremely disruptive to the air quality in the region, and the SH110 divided the village into two parts, thereby blocking the drains to Manimangalam lake and causing shortage of water for irrigation in the agricultural lands in those regions, resulting in the fallow lands. The fallow lands slowly transformed by real estate into residential and industrial areas, and the JK tyres factory on the edge of Malaipattu lake and the Adyar branch have influenced the region by a large number of migrating people from North India for labor, thus changing rural landscapes for residential development in the region.

6.4 Mahanyam Village

Mahanyam village is 37 m from MSL with population of 1750 in the Pedi plains of Malaipattu hills with a scrub forest of tropical dry deciduous trees. The forest spreads around 360 ha with lot of rich flora and fauna are seen in the forests. The forest edge is slowly degraded by the industry along its edge and the forest has been fragmented by the road in between fro the industrial usage, the forest is in slow succession stage with social forestry and community forestry development for the rural people livelihood in the region. The lakes in the catchment of Mahanyam reserve forest have two big lakes called Sitteri and Periyaeri/tank but the drains to the lakes were blocked and the land around the catchment area is encroached and left dry for half a year, so the irrigation for the nearby farmlands in the region is adequate and many areas are converted as fallow land and sold for real estate development as residential and industrial lands.

Mahanyam has a very old and famous Anjanayar temple with a lot of cultural rituals and festivals bringing thousands of people from surrounding to this region.

All this resulted, a magnificent rural landscape area on the urban outskirts is falling as a victim to urbanization, industrialization and transport developments and is losing its interest. The need to protect these high conservation potential zones is a strong need of time, as this rural area serves as a food production zone to meet the needs of urban areas with a large population and also acts as a buffer for high levels of pollution in this region.

7 Manimangalam Watershed Region

Issues in study site of Manimangalam, Malaipattu and Mahanyam villages.

Issues in the study site were seen in terms of issues found in land, water and forest which are the richest source of a region.

7.1 Land Use Change

The land use pattern in the Village Directory conforms to the pattern of classification of land use as recommended by the Ministry of Agriculture, Government of India. The ministry has recommended the maintenance of records of land use pattern under the following nine categories.

Land types classified in Manimangalam watershed by agricultural department

1. Wet agricultural lands (Nanseinilam)
2. Dry agricultural land (Punseinilam)

3. (a) Wasteland under cultivation—wet land (Nanseinilam)
   (b) Wasteland under cultivation—dry land (Punseinilam)
4. Forests
5. Grasslands and pasture lands
6. Non-agricultural lands other types (lakes/ponds/roads/buildings, etc.)

The land use is changing rapidly in the region due to the impacts of the urbanization, industrialization and transportation networks development.

According to land use land cover change comparing 2005, 2011 and 2018, there is an increase in built up rural by 1.97% and built up urban by 3.27% and fallow lands and barren lands increased due to reduced irrigation facilities and mining activities increased and water bodies reduced (Figs. 8, 9, 10).

8 Impacts of Urban Sprawl in the Hydrology, Forest and Land in the Peri-Urban Rural Landscapes in the Site

Hydrology

- Water channel blockage
- Siltation and sedimentation in lakes (less storage capacity)

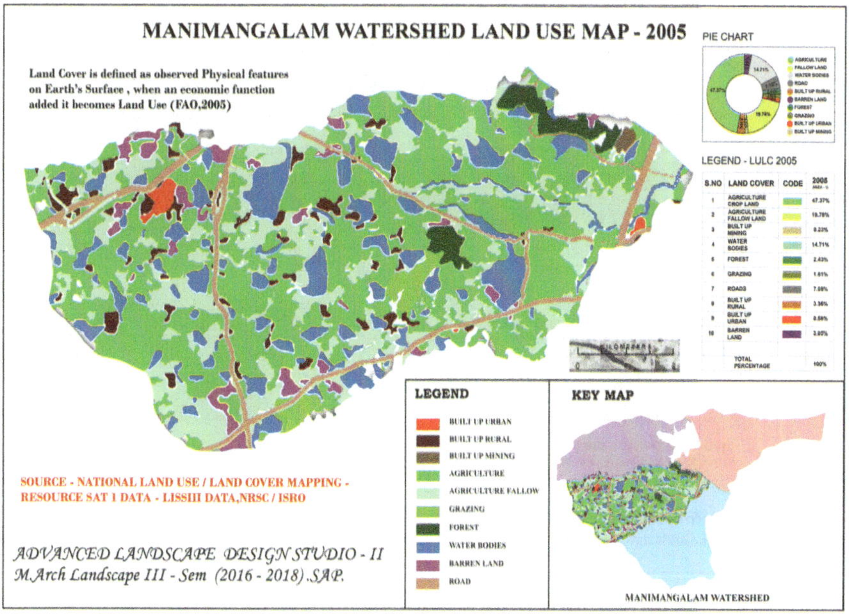

Fig. 8 Manimangalam watershed land use map 2005

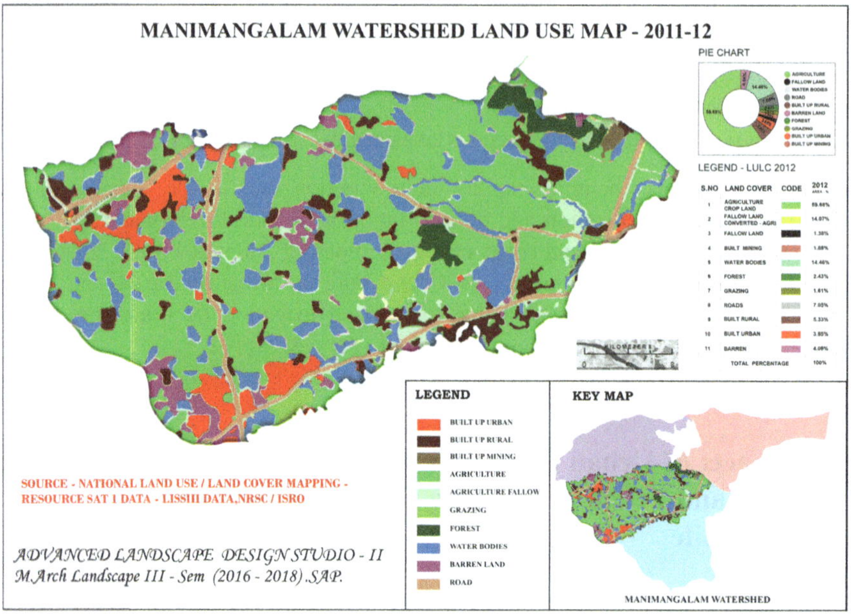

Fig. 9 Manimangalam watershed land use map 2012

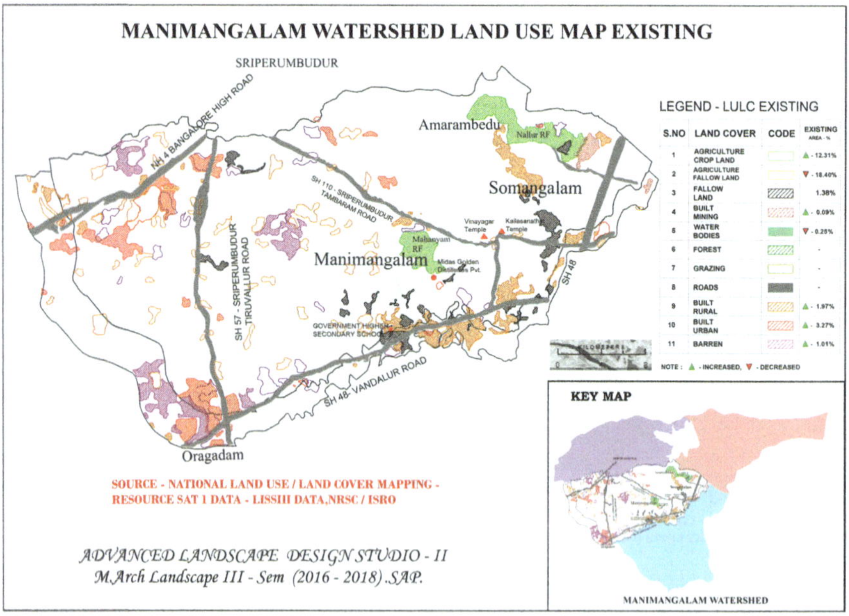

Fig. 10 Manimangalam watershed land use map existing

- Bund vegetation in less
- Obstructed linkage of system tanks in few areas
- Proper lining of bunds and channels is not found
- Encroachment in water bodies

Forest

- Open scrub due to soil erosion in hills
- Degraded vegetation along roadside of forest
- Boundary is not well protected
- Social forestry not aiming in native vegetation
- No restricted movement inside reserve forest
- Over grazing
- Watershed management is not integrated in forest area
- Biodiversity needs to be protected by proper buffering in corridors and boundaries
- Poaching of wildlife (boars entering farms)

Land

- Real estate development (more attraction for urban investments)
- Urban sprawl is the main reason of reduction of agricultural lands [3]
- Agricultural farm lands converted to fallow and waste lands
- Less infrastructure facilities lead to migration of people [4]
- Transit routes with no service roads (less safety for village people)
- Mining activities increasing pollution
- More industries (promote migratory population)

9 Recommendations for Conservation and Developmental Potential in the Study Site

Hydrology

- Clear obstruction in channels
- Desiltation of tanks
- Bund protection by vegetation
- Policies to control encroachment of water bodies with local people and government [5]
- Proper integration of watershed management and catchment area

Forest

- Restricted entry to forest till succession stage
- To avoid heavy vehicle entry for the industrial purpose
- To prevent ecological corridor disturbance
- Community participation in forest protection
- More native vegetation
- Biodiversity protection by buffering the core zone

Land Development

More attraction of urban people for country side living

- Land value is less so real estate development (air and water quality is good
- Eco-recreational development potential
- More cultural values and rich heritage values
- Popular temples and industries (attract more investment and employment)
- Policies for industrial developments without affecting the cultural values of village.

10 Visual Appraisal of Landscapes in the Study Site

Rich scenic values of the study site need to be protected from the impacts of urbanization (Fig. 11).

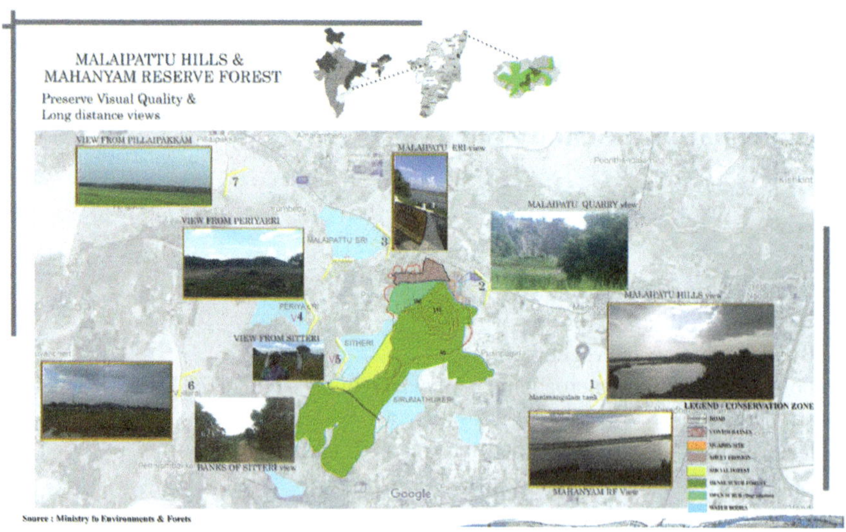

Fig. 11 High scenic value of rural landscapes in the study site

Strategies for watershed development activities

- Land development
- Water resource development
- Plantation activities
- Common property develops farm production
- Farm production
- Self-help groups and livelihood intervention for landless farmers

Guidelines for watershed development activities

Control of land uses in the watershed to prevent sedimentation of the reservoir and subsequently loss of storage, conversion of forest into agricultural land must be prevented. Understand and address natural flooding regime to decrease the loss of ecological values of the flood plains (agriculture) the regulation of the dam releases may be geared to the natural water and sediment demands.

11 Watershed Area Development—Recommendations

Catchment area

- Treatment of arable and non-arable land will arrest soil erosion and loss nutrient along with fertile soil (Fig. 12)
- Treatment of natural drainage channels, cleaning of feeder channels
- Scientific land use planning crops as per land capability tree crops/dry land horticulture increased tree cover, higher biomass production

Tank bed/offshore area

- Application of appropriate soil conservation measures and adoption of suitable cultivation practices
- Removal of encroachment and allowing free flow of water with the provision of silt traps, filter points, etc.
- Desilting of tanks will augment the storage capacity of tanks
- Tanks should be cleaned and made free of weeds during desilting operation
- Planting of bamboo and other trees in the foreshore will serve as filters and also be used by the poor as raw material for handcrafts
- Sluice, waste weir and operating system need to be repaired.

  Tank dam/bands should be strengthened by using excavated silt/other material. Pitching of interior wall with stone should be carried our wherever necessary.

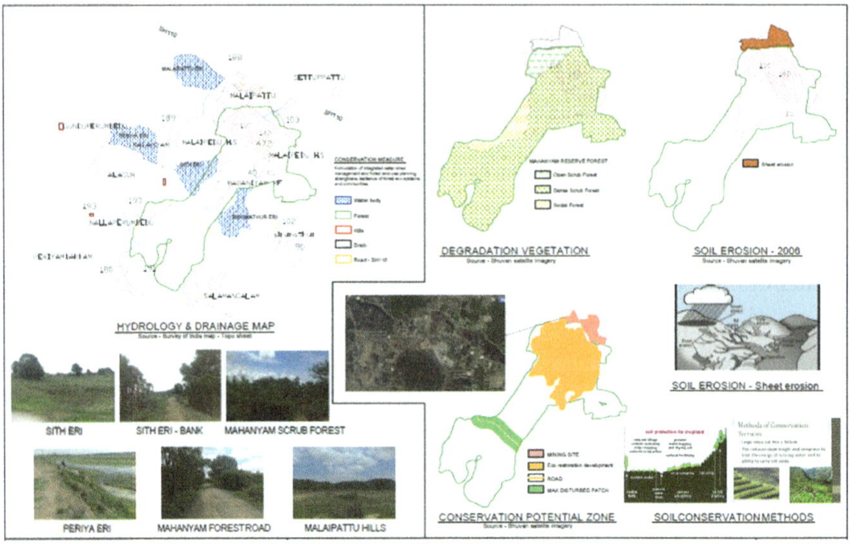

Fig. 12 Degradation of vegetation and hydrology due to urbanization in study site

Conservation strategy for in situ conservation of forest diversity

Strategy in the form of guideline represents technology natural resources base represents requirement of site, the reproductive patterns of the species, the diversity factors socio-cultural policy, incentives and capacity.

Factors affected forest development

Globalization, decentralization and privatization, changing demand for forest products and services from growing and often more urbanized population, awareness of role of forest in regulating climate, and other environmental services multifunctionality of forest need to be analyzed properly a shift timber centered to people centered forest management is necessary.

The program for strengthening resilience of forest ecosystems and communities includes the following

- Ecosystem-based vulnerability assessment
- Climate change adaptation planning
- Management of protection forests and protected areas
- Protection of existing forests
- Rehabilitation and conservation of vegetation
- Livelihood support
- Formation of integrated watershed management and forest land use plans [6].

Recommendations

Need for new policies and legal measures which can encourage more community participations in the forest management and can be provide them opportunities for forest-related livelihood, improvement of governance system and penalty and punishments for any sort of unlawful activity will conserve forest from illegal activities.

New Acts and Policy Recommendations

Acts and policy to regulate conservation of biological diversity and their sustainable use. The government can declare an area as "ECOLOGICALLY CRITICAL AREA" to restore/conserve biodiversity of the area. On consultation with local people, government can declare an area as "Heritage Site" for better management.

12 Conclusion

Urban expansion is rapidly expanding all over the world and the protection of rural landscapes in peri-urban areas has become mandatory. The need for time to preserve, rather than to act later on, potential visionaries for construction and urbanization needs to be controlled and unplanned urban development needs to be properly planned and a boundary of urban growth needs to be established.

Urban growth boundary needs to define the boundaries of urban areas and to maintain and protect the peri-urban areas and rural fringes of urbanization, transport and industrialization.

Future generation needs to understand the importance of peri-urban and rural areas and what ecosystem services it can provide and can reduce pollution and serve as a carbon capture for urban and industrial areas, and the productive landscapes also serve to meet growing demands, and the high scenic value and visual quality of rural landscapes need to be preserved.

Agriculture protection zones, vegetative belt protection, water bodies protection and soil conservation through various measures and farm land planning and management should be done. Transparency and people-friendly system for regulation of land, water and other natural resources Economic zone development swallowing the rich ecozone to be monitored and controlled through rules and regulations (value the land assessing the environmental value).

Government aids and policies to conserve rural area like protected rural zone should be framed using various stakeholder participation.

Community engagement in conservation will play a major role so it should be encouraged.

The study site is one such crucial site in the highly urbanized region of south Chennai and the industrial growth hub of Sriperambadur, and the growth and development of both sides is expanding at a rapid pace, with rural landscapes slowly falling under the strain of urbanization and losing value, and rural landscapes slowly declining and losing their identity.

Landscape strategy and planned urban projects need to include a boundary for urban growth that will act as a buffer and green belt to protect suburban and rural areas and the same way that urban populations will have a breathing space for leisure and relaxation and a green space for future generation experience and to reduce the impact of urbanization.

Policies and programs for the preservation of rural landscapes through a proper urban growth boundary will ensure the retention of parcels of land left over, which are very important for the growth and development of urban areas and will also reduce the impacts of urbanization and industrialization.

References

1. Christopher Bryant (2009) Impacts of urbanization on rural land use. Department of Geography, University of Montreal, Montreal, Canada, Eolss publisher Co. Ltd, Oxford, UK
2. AE Cengiz (2013) Impacts of improper land uses in cities on natural environment and ecological landscape planning. http://dx.doi.org/10.5772/55755© 2013 Cengiz, licensee InTech
3. Padmanaban R, Bhowmik AK, Cabral P, Zamyatin A, Almegdadi O, Wang S (2017) Modelling urban sprawl using remotely sensed data: a case study of Chennai City, Tamilnadu. Entropy 19(4):163
4. Sati VP, Deng W, Lu Y, Zhang S, Wan J, Song X (2017) Urbanization and its impact on rural livelihoods: a study of Xichang City administration, Sichuan Province, China. Chinese J Environ Studies. https://doi.org/10.1142/S2345748117500282
5. Bengston DN (2004) Policies for managing urban growth and landscape change: a key of conservation in the 21st century. In: Symposium, society for conservation biology 2004. Annual meeting, Columbia University, New York, 30 July–2 Aug 2004
6. Wandl A, Magoni M (2017) Sustainable planning of peri urban areas: introduction to the special issue, planning practice and research. 32(1):1–3, http://dx.doi.org/10.1080/02697459.2017.1264191

Sustainable Urban Development

A Review on Plastic Waste Assessment and Its Potential Use as Building Construction Material

Akhilesh Kumar and Avlokita Agrawal

Abstract Plastic waste has become a severe threat to the environment globally, and unmanaged disposal practices have made it more challenging. A massive amount of various types of plastic are piling up at landfill sites, chocking drainages, which has now entered into our food chain as animals and fishes are eating plastics. Ocean and their aquatic life are also getting suffered from plastic waste as it takes millions of years to degrade naturally. Open burning of the plastic waste courses major health issues like the respiratory problem, irritation, and even cancer in some of the cases. Recent research depicts how plastic waste is also one of the significant reasons for global warming and the rapid extinct of indigenous species. Other studies showcase how virgin and raw materials have been extracted to fulfil the demands of the construction industries. On the other hand, with the unique interventions, some of the cities in the various regions have developed strategies to fight with the problem and material demand—as they are utilising the different type of plastic waste into various kind of building and construction materials. Hence, studies show significantly improved impact on the environment. The involvement of the informal sector could help by contributing to developing the economy and livable environment—that is how the principal amount of resourceful plastic waste could prevent to the reach at dumpsites and became a resource for construction material.

Keywords Alternative technique · Reuse · Conservation · Plastic waste · 3D-printing · Sustainable

---

A. Kumar (✉) · A. Agrawal
Department of Architecture and Planning, Indian Institute of Technology, Roorkee, Uttarakhand, India
e-mail: akumar1@ar.iitr.ac.in

A. Agrawal
e-mail: avlokita@ar.iitr.ac.in

K. Thirumaran et al. (eds.), *Sustainable Urban Architecture*,
Lecture Notes in Civil Engineering 114,
https://doi.org/10.1007/978-981-15-9585-1_3

1 Introduction

Plastic is ubiquitous, that is visibly the backbone of globalisation. Plastic is everywhere as much as we see and found it is durable, strong, easy to mould in any shape and waterproof nature. In spite, all plastic has become the biggest threat to the environment. Plastic was invented in 1839 first in the form of polystyrene by 'Eduard Simon' than in 1862—'Parkesine' (an early type of plastic) was invented and demonstrated publically by 'Alexander Parkesto' claims to protect the environment through his invention. As lots of trees were being cut for fulfil, the demand of the paper packaging eventually now has created a disasters situation of all species [8].

As per the recent 'United Nations Environment Programme' report (UNEP), the world produces 400 million tonnes of single-use plastic waste (SUPW) yearly, which is about 47% of the total plastic waste also expected to reach 500 million tonnes by 2030 referring Fig. 1. It has estimated that only 9–10% of plastic wastes are getting recycled worldwide [25], rest goes for open burning and into the ocean via river through city drains. Haphazard economical and industrials growth in the urban area, more consumer-oriented lifestyle and continuous changing trends leads us in this worse situation and forced to generate and use more plastics and packing.

Plastic is also causing cancer; open dumping and burning of the plastic are found a common easy practice showcased by most of the municipalities without any much-intended future goals.

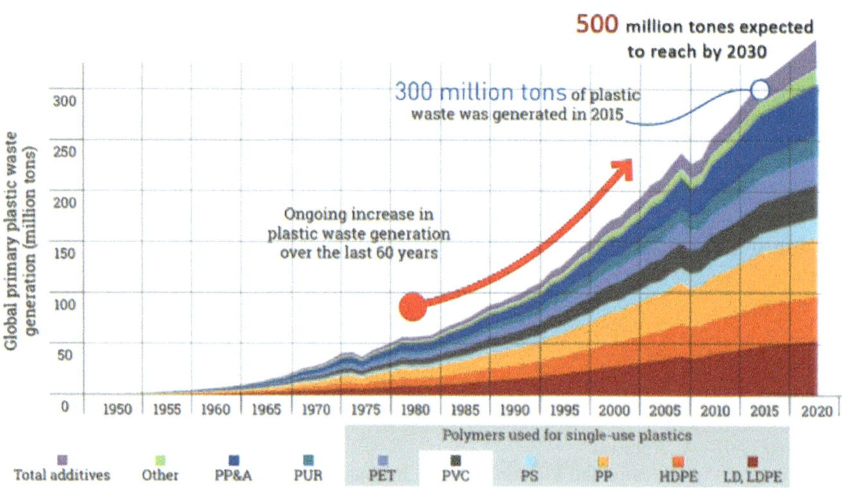

Fig. 1 World primary plastic waste generation from the year 1950–2020 [9, 13]

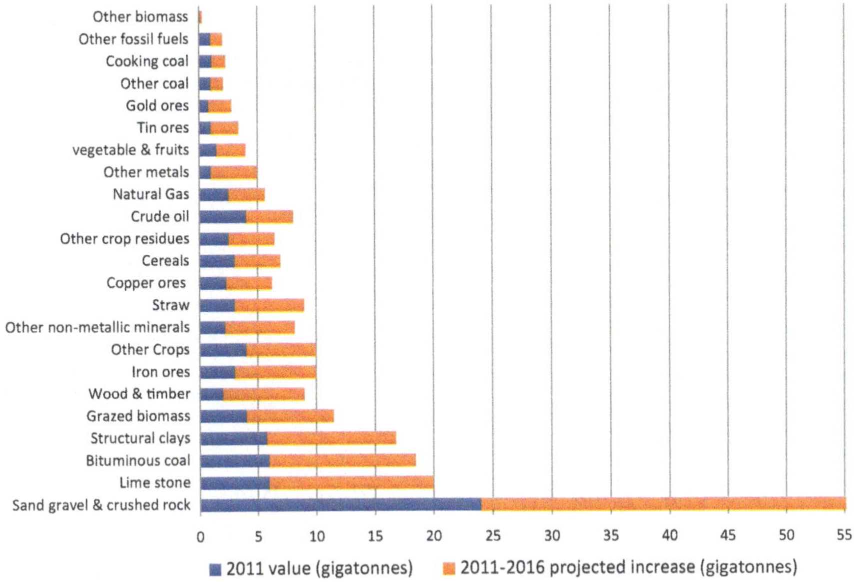

Fig. 2 Total materials domination use in (2011 and 2060) for construction [15]

Impacts of mismanaged single-use and micro-plastics

(a) A threat to the economy: tourism, fisheries, agriculture
(b) Implications for human health: food chain contamination, toxic fumes on burning, blockage of drains to flood, contamination of water resources
(c) Impact on the environment: loss of biodiversity, air, and land pollution, sea pollution, global warming.

What are the opportunities we can gain through plastic waste and their reuse? On the other hand, plastic has showcased the tremendous opportunity as construction materials, and the study has presented that plastic waste could become a significant alternative to the virgin materials as it has the flexible, durable, and sustain more extended periods. Referring to Figs. 1 and 2 represents how virgin materials may rise to the existing to fulfil the global infrastructure development demands in the future. However, this is additional pressure on the materials market.

1.1 Definitions and Facts on Plastic Waste

Definition of plastics—A synthetic polymer and substitute of natural materials has become an essentials aspect of our life. Referring to Fig. 3 shows how global plastic production by the industrial sector is been dominating the production and promoting plastic consumer-oriented habits.

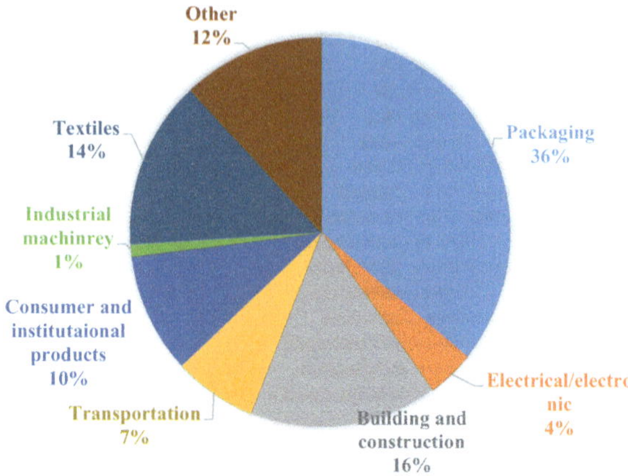

Fig. 3 Global plastic production by industrial sector, 2015 [34]

1.2 Facts and Media Reporting on Plastic Waste

- The factual report published in Teri India says micro-plastic and plastic debris affects at least '267 species' globally, including 86% of all 'sea turtle species', 44% of all 'seabird species', and 43% of all 'marine mammal species'.
- An enormous amount of toxic heavy metals like copper, zinc, lead, and cadmium recovered from plastic wastes from seashores hurt the coastal ecosystems.
- Several GHG gases are emitted from the landfills. As CO_2 and methane etc. constitute 90–98%.
- Plastic waste to overweight fish weight by 2035. Plastic bags and pieces of styrofoam remain intact until 1000 years and even more [9].
- The world's plastic production uses about 8% of the world's oil production, 4% as a feedstock and 4% during the processes of manufacturing [29].

1.3 Current Plastic Waste Status, Practices and Its Management

As per the centre pollution control board, India estimated the collection efficiency of plastic waste and other mixed waste as 80.28% in 2014, out of that only 28.4% was treated. The remaining collected waste quantities were disposed of in landfills or open dumps. Eventually, 60% of collected plastic only goes to recycling; about 43% of total single-use plastic waste comes from various packaging waste.

The recent study conducted by the 'Pandit DeendayalUpadhyaySmritiManch' represents, more than 121 thousand metric tonnes of plastic waste are being imported in India by the recycling companies from various more than 25 countries, out of which 55,000 metric tonnes of plastic are imported from Bangladesh and

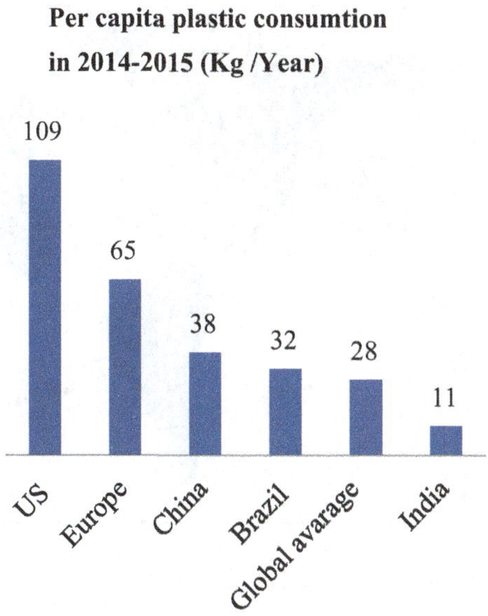

Fig. 4 Global trend in plastic waste. *Source* FICCI, [27]

Pakistan. Delhi along imported about 19,000 metric tonnes of plastic to fulfils the demand [23].

Referring to Fig. 4, the global trend in plastic waste as Indian consumes 11 kg plastic every year compared to an average American 109 kg [28]. As Fig. 5 indicates, the top regional municipal plastic waste generator Mt/Year in the world. The Asia continent is becoming a massive plastic waste generator. Referring to Fig. 6 represents the concentrations of plastic debris in surface waters of the global ocean. Seeing to Figs. 7 and 8 indicates how plastic waste been increased in India with time and comparison with other countries.

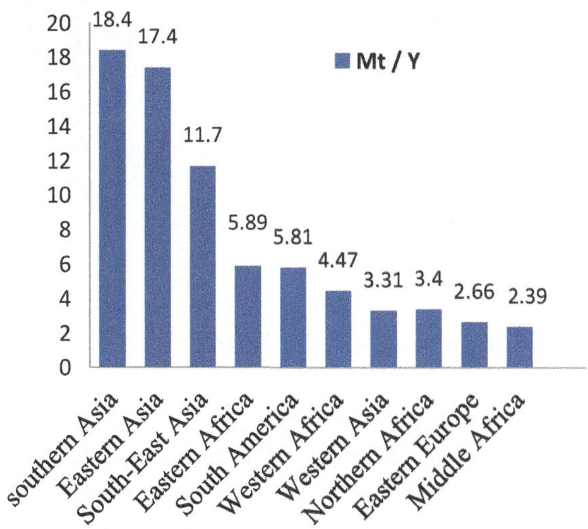

Fig. 5 Top regional of municipal plastic waste in MT/Year for 2015 [18]

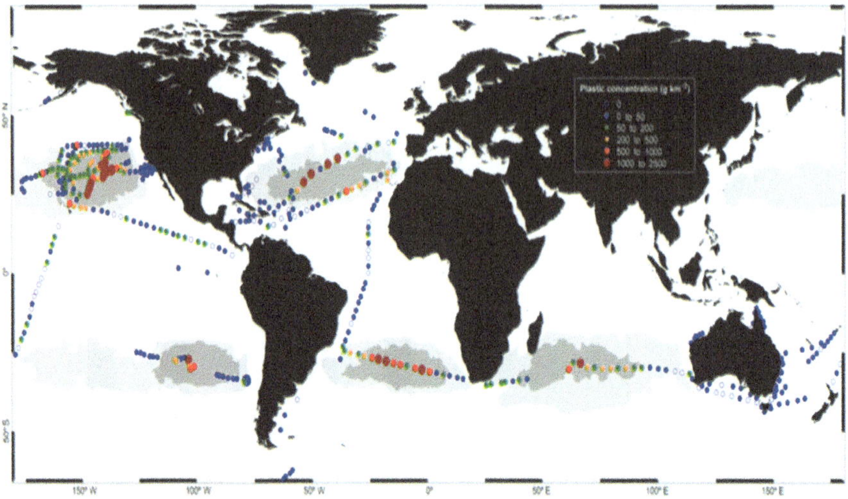

Fig. 6 Plastic debris concentrations in the surface waters of the global ocean [10]

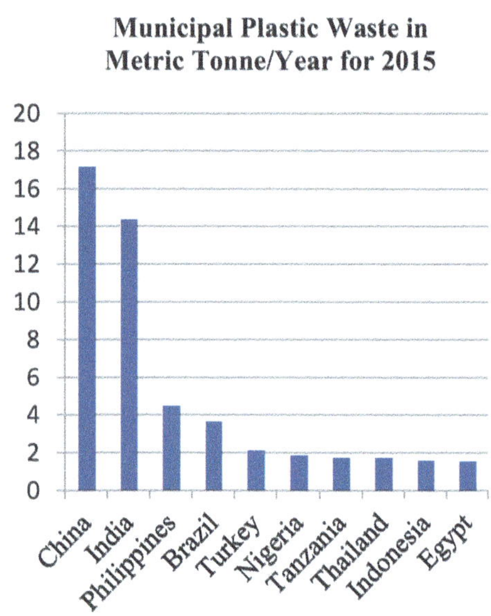

Fig. 7 Top countries of municipal plastic waste in metric tonne/year for 2015 [18]

2 Significance of the Study

The shortfall of the resources building material is putting additional pressure on virgin materials. A considerable amount of Greenhouse gases emission increases during the pre and post-construction processes. The massive resourceful plastic

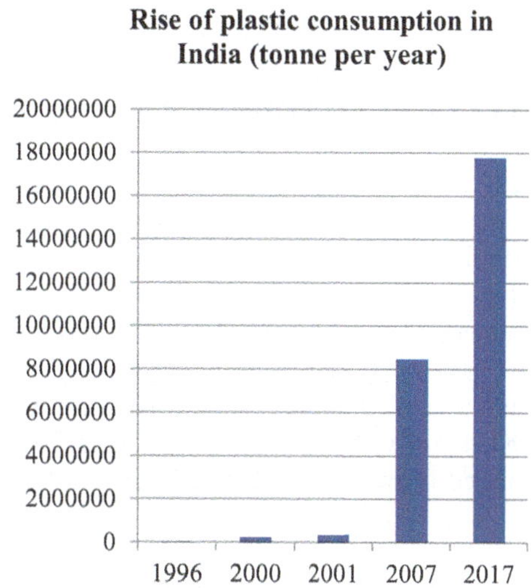

Fig. 8 Rise of plastic consumption in India [27]

waste piling up on landfill sites and ocean surface create an un-reversible environmental impact. Social and economic upliftment of poorly involved waste pickers through the new emergence of waste plastic into building construction material.

3 Materials and Methods

In this study, the Journal articles/papers on plastic and their applications have been reviewed, 60% were directly on Plastic waste, and about 30% were indirect advice. The available literature on the web talks about its physical and chemical properties, specifically for industrial purposes. The bifurcation of the study is on the following points;

(a) Selection of literature on plastic waste.
(b) Best cases significance on plastic and their application in building and construction.
(c) Availability of the materials in the context of South Asia.
(d) Feasibility of the plastic waste material in a different part of the building.
(e) The result on the bases of assessment of the properties and applications of materials.

3.1 Best Practices of Plastic Waste

3.1.1 Case 01: Plastic for Low Cost, Efficient Building, and Roads

The study conducted in India revels that recycled plastic waste in building construction works efficiently in the earthquake scenario due to lower dead load (182–195 kg/m^3) in beam-column and other structural parts. As compared to conventional mixed gravel concrete (2400 kg/m^3), red-brick wall (1920 kg/m^3) and cement density (1500 kg/m^3). The cost of the construction can reduce up to 1/3 of the total cost. The environmental analysis shows utilising of recycled plastic waste lumps can help to prevent environment as production of 1 tonne of cement releases 900 kg of CO_2. Hence, study shows 30–40 thousand kilos of plastic can be consumed in small scale house limits plastic to reach to the landfills [16].

3.1.2 Case 02: Involvement of an Informal Sector-Circular Economy

The city of Pune, India, has involved about 3,000 swatch workers in the plastic collection to develop a closed-loop/circular economy. About 166 tonnes of plastic waste have upcycled into various useful products that saved CO_2 emission of 50,000 tonnes annually, equivalent to 10,423 passenger cars on the road. Efficient management of the formal and informal waste sector helps municipally set an example of Circular economy generation [21]. Referring to Fig. 9 for the process of circular economy development.

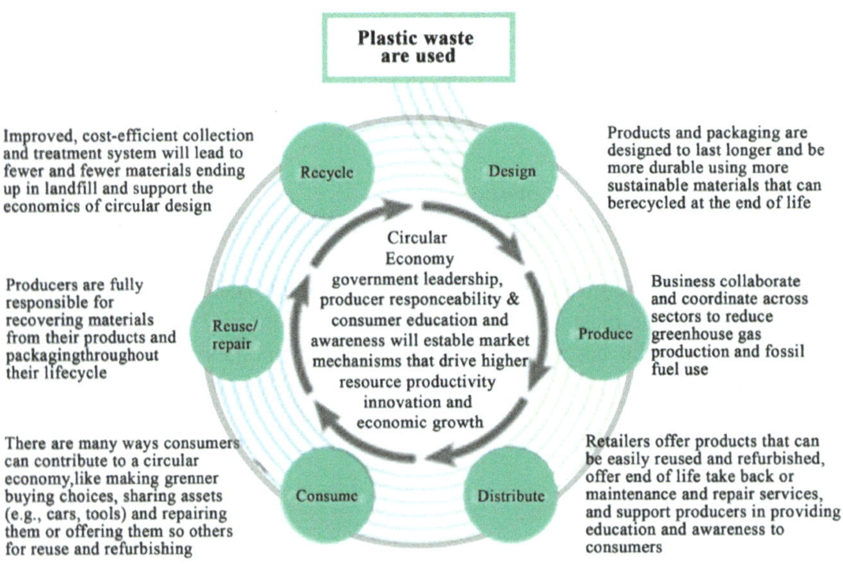

Fig. 9 Circular economy development process [9]

4 Plastic Waste at the Different Part of the Building Construction

4.1 Plastic Waste as Rough Materials: Aggregate, Cement, and Sand

(a) Plastic waste as a replacement of the cement for making tiles: 50% of the plastic waste with the weight of sand shows the excellent result of predation tile, which is also found durable cheaper as compare to cement tiles [1].
(b) The plastic polymer in a mortar, repair work, marine construction, precast products, plastic aggregates, and partial replacement of sand found suitable [7].
(c) The utilisation of 'Plastic Bottle Waste' crushed in concrete as aggregates found remarkable results to its inflexible and durable strengths [5].
(d) The waste plastic materials for road construction high-density polyethene, polypropylene has shown auspicious results in the road construction with better cracking resistance, rutting under different conditions [6].

4.2 Structural Elements Beam and Columns

(a) The plastic fibres were added up to 3.0% found improved strengths in compression of concrete determined after 28 days of its curing period and compared with 'control concrete'. The experiment shows an increase in compressive strength by 12% [5].
(b) E-plastic waste in concrete found the yield of concrete reduces when E-waste is adopted as a partial or complete replacement material for sand [4].

4.3 Plastic Waste as Filler and Soil Stabiliser

(a) As per the several tests conducted, 'California Bearing quantitative' relation (CBR) by the researcher shows how plastic waste strips along with the soil help to improve the shear strength and bearing capacity of the soil and plastic waste resin a filling component [11].
(b) The higher compressive strength has obtained for the compressed earth bricks containing 1% waste plastic of sizes lesser than '6.3 mm', and its 'compression strength' and flexible amounted to a 244.4% increment. The CEB samples stabilised with chopped 'plastic waste chips' containing 1% plastic waste of sizes lesser than 6.3 mm also had a minimal erosion rate [3].

4.4 Direct Uses of PET Bottles

(a) A solar bulb from pet water and bleach emits enough light could be the cheapest light source for remote areas, its insoluble nature for more than 300 years makes it suitable for construction [33].
(b) Chipped recycled aggregate filled PET plastic bottles offer comparable 'compressive strength' as compared to a conventional clay/red brick [17].

4.5 Plastic Brick Blocks

(a) Compressive strength of plastic sand brick is 5.6 N/mm^2 at the compressive load of 96 KN. The study concludes that the plastic-sand bricks are found useful for the construction comparing with fly-ash bricks and third class clay bricks/blocks [19].
(b) 'Sand plastic brick' has a lower rate of water absorption, bulk density, and apparent porosity when it is compared with the standard clay bricks and higher compressive strength [30].

4.6 Panel Wall/Ceiling

(a) The block of air-filled bottles could be utilised for building construction as either a lightweight partition walls for multi-story building blocks or even for a bearing wall roof slab, which has shown a factor of safety to 5.8 in a single-story building [20].
(b) Study envisaged the possible uses of glass mixed with plastic waste in the plastering material, as reinforcement, the results represent the 'flexural testing ability' of all beams with plastic fibres confirms high tensile strength and lightweight properties [24].

4.7 Furniture and Artefacts

The 'plastic-wood' its applications for making various types of furniture. Utilising recycled plastic to developing polywood, the recycled plastic can be employed as a resin to produce artefacts and other moulded furniture.

4.8 Roof, Floor, and Pavement Tiles

Plastic waste has shown the significant potentials to develop the pavement tile/ block with equal and almost similar compressive strengths at the industrial scale and also reduces up to 600 kg of plastic waste to build one thousand blocks and also be used with sand by its 50% weight in non-traffic and light traffic roads tiles/ blocks [14].

4.9 Binding and Pasting

(a) Pet bottle gum-plastic waste as an alternative binder for paving blocks production, pet bottle furniture fixing with hot dryer process.
(b) Thermocol/styrofoam has showcased that 'petrol and toluene' are able to give good quality of adhesive than that of other solvents. The technique is cheaper as it can utilise the waste styrofoam. The solvents mixture of 'acetone' and 'petrol' with thermocol gives adhesive [26].
(c) In road construction, an appropriate amount of the added PET waste is found suitable when 6% by weight of bitumen is mixed. That also gives a maximum level of stability and high Marshall Quotient [2].

4.10 Sound Insulation and Thermal Comfort

Utilising the plastic waste as thermal insulation has shown the positive results to the environmental protection point of view. Waste plastic wool and recycled form of the polyester fibres (RPET) for construction and building industry applications suitable for sound and thermal insulation properties are confirmed [22].

4.11 Petro/Diesel Oil, Lubricate, and Gas

In the process of pyrolysis, the plastic waste and 'Fischer-Tropsch' (FT) wax to lube range molecules and diesel base gases and gasoline products from plastic wastes [32].

4.12 *Clothes and Textiles Fibres*

Pet bottles have shown a tremendous opportunity to develop clothes and textile fibres. A company in NewYork city used 4 billion bottles for making clothes, shoes, T-shirt and graduate grows.

4.13 *Advanced 3D-Printing: Plastic Waste-Based Practices*

3D is no longer a future technology, and it has utilised by all around the world. We can now able to print whole houses within a short time. The plastic from petroleum-based and e-waste has the potentials to make the filaments for 3D printing and 30% lesser carbon emission to the environment [12]. NASA, USA, has demonstrated that plastic waste in their all upcoming space mission would be to reuse–recycle through 3D-printing process into the required tools and pieces of equipment [31].

5 Assessment and Properties of Plastic Wastes as Construction Materials

See the Table 1.

6 Results and Discussion

To overcome problems like scarcity of material, and increase cost, how recycled plastic could be used at an industrial scale. There is an urgent need for efficient management in the plastic waste sector through the formal involvement socio-economic section in waste management practices, with private and government agencies such as stakeholders & self-help groups and integrated monitoring agencies. Also, to ensure collecting resourceful materials, segregation, pre-processing recycling, and reuse. About 90–100% of the total plastic waste can be removed from the landfill site and could be utilised as a resourceful material for constructing a building and their different parts. The holistic and integrated approach towards plastic waste is required to establish at the centralised and decentralised level. Referring to Table 1 and Fig. 10 represents material property assessment and utilisation of the various plastic wastes at different building profiles.

Table 1 Plastic waste as building material and assessment of properties

Properties	PETE: is terephthalate	HDPE: high density polyethylene	PVC: polyvinyle chloride	LDPE: low density polyethylene	PP: polypropylene	PS: polystyrene	Others plastics
Structural	1, 3, 5, 7, 10, 11	2, 4, 5, 6, 8, 11	1, 5, 7, 9, 10	2, 4, 5, 6, 8, 11, 12	1, 2, 3, 8, 10, 11, 12	1, 5, 7, 8, 10, 11	5, 6, 10
Thermal comfort	Ext	Ext	Ext	Ext	Ext	Ext	Mod
Temperature resistance	Mod	Ext	Mod	Mod	Mod	Mod	Ext
Water resistance	Ext	Ext	Ext	Ext	Ext	Ext	Ext
Sound insulation	Ext	Ext	Ext	Ext	Ext	Ext	Ext
Buildability	Ext	Ext	Mod	Ext	Mod	Mod	Mod
Cost-effective	Yes	Yes	Dpn	Yes	Yes	Yes	Yes

Structure properties: 1-Lightweight, 2-Flexible strength, 3-Melting, 4-Tensile strength, 5-Corrosion resistance, 6-Load bearing, 7-Fiber content, 8-Modulus elasticity, 9-Admixture, 10-Pozzolanic, 11-Reduce pollution, 12-Shear forces

Physical Properties: {thermal comfort, water resistance, buildability, cost-effective (yes/no), eco-friendly, thermal insulation, sound insulation}. Scale-(Ext)-Excellent, (Mod)-Moderate, (Pr)-Poor, (Dpn)-Depends on its compositions

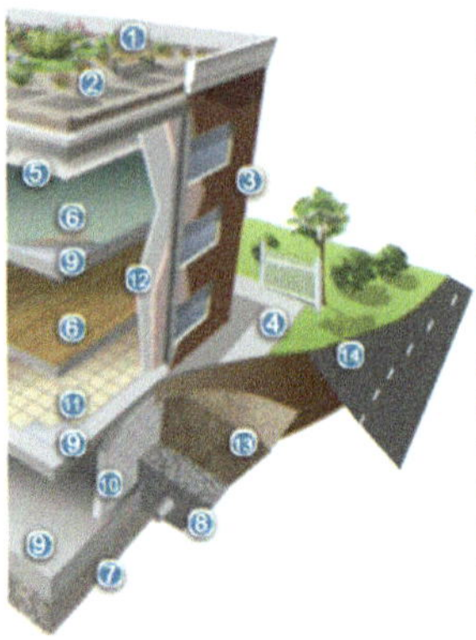

Fig. 10 Plastic waste utilisation at different parts of the building

7 Conclusion

This study is an attempt to overview the plastic waste status, current problems globally, and their potential application as building construction materials, to raise sustainability. It will be more expensive to treat and reuse of the plastic waste in the future as compared to today, huge job opportunity, boost in circular material demands, social and economic upliftment of involved sectors. This paper has reviewed and assessed the various possible literature, research articles, books, reports, best practices, case studies, and government documents on plastic waste uses to develop the methodology which showcases the plastic wastes as an alternative building material utilisation and advanced 3D-printing techniques, that even opens a new window for creating a sustainable and livable environment for all species and also envisaged the scopes for further investigation and research.

References

1. Abdul A, Momin R (2014) Plastic floor tiles. ICCE-Elsevier, pp. 282–290. Available at: https://www.researchgate.net/publication/322789874_Plastic_Floor_Tiles
2. Ahmadinia E et al (2011) Using waste plastic bottles as additive for stone mastic asphalt. Mater Des 32(10):4844–4849. https://doi.org/10.1016/j.matdes.2011.06.016
3. Akinwumi II, Domo-Spiff AH, Salami A (2019) Marine plastic pollution and affordable housing challenge: shredded waste plastic stabilized soil for producing compressed earth bricks. Case Stud Constr Mater 11:e00241. https://doi.org/10.1016/j.cscm.2019.e00241
4. Alagusankareswari K et al (2016) An experimental study on E-waste concrete. Indian J Sci Technol 9(2):3–7. https://doi.org/10.17485/ijst/2016/v9i2/86345
5. Alsadey S (2016) Utilization of plastic bottle waste in concrete. J Biol Environ Eng 1(1):1–5. Available at: http://www.aiscience.org/journal/jbee

6. Appiah JK, Berko-Boateng VN, Tagbor TA (2017) Use of waste plastic materials for road construction in Ghana. Case Stud Constr Mater 6:1–7. https://doi.org/10.1016/j.cscm.2016.11.001
7. Awasthi AK, Shivashankar M, Majumder S (2017) Plastic solid waste utilization technologies: a review. IOP Conf Ser Mater Sci Eng 263(2). https://doi.org/10.1088/1757-899x/263/2/022024
8. Bellis M (2019) A brief history of the invention of plastics. ThoughtCo. Available at: https://www.thoughtco.com/history-of-plastics-1992322. Accessed on 8 Nov 2019
9. Claudia Giacovelli (2018) UNEP (2018), Single-use plastics: a roadmap for sustainability. In: Tara C (ed). Available at: https://wedocs.unep.org/bitstream/handle/20.500.11822/25496/singleUsePlastic_sustainability.pdf?sequence=1&isAllowed=y
10. Cozar A et al (2014) Plastic debris in the open ocean. Proc Natl Acad Sci 111(28):10239–10244. https://doi.org/10.1073/pnas.1314705111
11. DhonSA, Borkar DB (2018) Stabilization of soil by using plastic wastes. IJSRD—Int J Sci Res Dev 6(09):2321–0613. Available at: https://www.academia.edu/37924440/Stabilization_of_Soil_by_Plastic_Waste_IJSRDV6I90136
12. Gaikwad V et al (2018) Transformation of E-Waste plastics into sustainable filaments for 3D printing. ACS Sustain Chem Eng 6(11):14432–14440. https://doi.org/10.1021/acssuschemeng.8b03105
13. Geyer R, Jambeck JR, Law KL (2017) Production, use, and fate of all plastics ever made. Sci Adv 3(7). https://doi.org/10.1126/sciadv.1700782
14. Ghuge J et al (2019) Utilization of waste plastic in manufacturing of paver blocks. Int Res J Eng Technol 06(04):1967–1970. Available at: www.irjet.net
15. 'Global Material Resources Outlook to 2060' (2019) Global material resources outlook to 2060. https://doi.org/10.1787/9789264307452-en
16. Gulhane S, Gulhane S (2017) Analysis of housing structures made from recycled plastic. IRA-Int J Technol Eng (ISSN 2455-4480) 7(2 (S)):45. https://doi.org/10.21013/jte.icsesd201705
17. Lalzarliana Paihte P, Lalngaihawma AC, Saini G (2019) Recycled aggregate filled waste plastic bottles as a replacement of bricks. Mater Today Proc 15:663–668. https://doi.org/10.1016/j.matpr.2019.04.135
18. Lebreton L, Andrady A (2019) Future scenarios of global plastic waste generation and disposal. Palgrave Commun 5(1):1–11. https://doi.org/10.1057/s41599-018-0212-7
19. Manjarekar PAS (2017) Utilization of plastic waste in foundry sand bricks. Int J Res Appl Sci Eng Technol V(IV):977–982. https://doi.org/10.22214/ijraset.2017.4178
20. Mansour AMH, Ali SA (2015) Reusing waste plastic bottles as an alternative sustainable building material. Energy Sustain Dev 24:79–85. https://doi.org/10.1016/j.esd.2014.11.001
21. Moora H et al (2019) Innovative partnerships with informal workers to recover plastic waste, in an inclusive circular economy approach. Pune. Available at: https://www.unescap.org/sites/default/files/Closing The Loop_Pune%2C India Case Study.pdf. Accessed on 4 Dec 2019
22. Patnaik A et al (2015) Thermal and sound insulation materials from waste wool and recycled polyester fibers and their biodegradation studies. Energy Buildings 92(April):161–169. https://doi.org/10.1016/j.enbuild.2015.01.056
23. PTI (2019) Over 1,21,000 MT plastic waste being "slyly" imported in India from 25 countries: study—the economic times. Econ Times. Available at: https://economictimes.indiatimes.com/news/politics-and-nation/over-121000-mt-plastic-waste-being-slyly-imported-in-india-from-25-countries-study/articleshow/70462997.cms?from=mdr. Accessed on 6 Nov 2019
24. Salim K et al (2019) Reinforcement of building plaster by waste plastic and glass. Procedia Struct Integrity 17:170–176. https://doi.org/10.1016/j.prostr.2019.08.023
25. Santhosh S, Shrivastav R (2019) Plastic waste management: what can India learn from other countries. DownToEarth, 3 November. Available at: https://www.downtoearth.org.in/blog/waste/plastic-waste-management-what-can-india-learn-from-other-countries-67048. Accessed on 8 Nov 2019

26. Selukar NB, Lande CV, Ingole CG (2014) Waste thermocol to adhesive for better environment. Int J Innovative Res Adv Eng (IJIRAE) 1(6):2163–2349
27. Sharma S, Mallubhotla S (2019) Plastic waste management practices—issues, solutions and case studies, ministry of housing and Urban Affairs. GOI, New Delhi. https://doi.org/10.1201/9780429059247-7
28. Shreeshan Venkatesh IK (2018) An Indian consumes 11 kg plastic every year and an average American 109 kg. DownToEarth, 6 June. Available at: https://www.downtoearth.org.in/news/waste/an-indian-consumes-11-kg-plastic-every-year-and-an-average-american-109-kg-60745. Accessed on 6 Nov 2019
29. Siddique R, Khatib J, Kaur I (2008) Use of recycled plastic in concrete: a review. Waste Manag 28(10):1835–1852. https://doi.org/10.1016/j.wasman.2007.09.011
30. Singh LB et al (2017) Manufacturing bricks from sand and waste plastics. Int J Eng Technol Manag Appl Sci www.ijetmas.com 5(3):426–428. Available at: www.ijetmas.com
31. Skrabania L (2018) Sustainable 3D printing—with plastic waste and sugar beets, development| RESET.org. Available at: https://en.reset.org/blog/sustainable-3d-printing-plastic-waste-and-sugar-beets-06252018. Accessed on 8 Nov 2019
32. Thahir R et al (2019) Production of liquid fuel from plastic waste using integrated pyrolysis method with refinery distillation bubble cap plate column. Energy Rep 5:70–77. https://doi.org/10.1016/j.egyr.2018.11.004
33. Wang C et al (2015) Critical view on daylighting through solar bottle bulb. Build Res J 61(2):115–128. https://doi.org/10.2478/brj-2014-0009
34. Zink T, Geyer R, Startz R (2018) Toward estimating displaced primary production from recycling: a case study of U.S. aluminum. J Ind Ecol 22(2):314–326. https://doi.org/10.1111/jiec.12557

The Sustainable City Intercellular Approach, Study of Urban Fabric Pattern, a Case Study of Chennai City

K. Madhivadhani

Abstract As per the census data 2011, more number of people in Tamil Nadu shifted from rural to urban area compared to the other states. The level of urbanisation in Tamil Nadu increases nearly 14.3% over the last two decades. It results in emerging informal settlements, lack of infrastructure facilities and decline of economic growth, etc. as a whole it reflects on the chaotic spatial pattern. The sustainable urban fabric pattern is an intellectual strategy and a sequential implementation process of integrating rural and urban areas. It is important in city planning to attain the balanced development of the region to fulfil the need of the backward villages. Intercellular approach reduces the conflicts and competition for resources between cities in a region. The establishment of an efficient relationship between living place, work place and social service and recreational area so as to enhance the quality life of the metropolitan population are the main task in the process of intercellular planning. This research is an attempt to achieve sustainable environment with the help of urban planning. Further this article deals with the development strategy for future urban growth through urban planning models.

Keywords City intercellular model · Cumulative functional index · Industrial development · Spatial distribution

1 Introduction

Urbanisation is an irreversible process accelerated with the explosion of population and technological transformations. As the same time urbanisation may as well associated with chaos and socio-economic disorders and deprivations and need to be channelled and regulated in proper direction. Urbanisation is an uninterrupted continual process of shifting rural population from agricultural sector to manu-

K. Madhivadhani (✉)
Department of Planning, School of Architecture and Planning, Anna University, Guindy, Chennai, India
e-mail: vadhani27@gmail.com

K. Thirumaran et al. (eds.), *Sustainable Urban Architecture*,
Lecture Notes in Civil Engineering 114,
https://doi.org/10.1007/978-981-15-9585-1_4

facturing and service sectors. It is a way to modernise the world through transformation. In Indian context, it plays a major role in economic development by generating income from industrial and other service sectors. But on the other side, emerging urban agglomeration results in the uncontrolled development, that ends in urban sprawl and results in depletion of natural resources. The bilateral nature of urbanisation always creates positive and negative impact on the spatial pattern. Intercellular approach is much needed one for the urban environment to sustain its diversified development over a period of time. The sustainable urban fabric pattern is an integration of various sectoral aspects such as linking of transit networks, promotion of Industries and enhancing of the city image. It is an imaginative simulation approach, to have alternatives and to be equipped with various incentives. This approach is important to examine the over exploitation and squander of natural resources from the rapid urbanisation by rejuvenating the resources for the safe and quality human life through spatial planning.

2 Urbanisation in Chennai

Chennai being the capital of Tamil Nadu is the major commercial centre and one of the largest industrial hubs in the entire state. The Corporation of Chennai expanded by taking into itself 42 local bodies, which includes nine municipalities, eight town panchayatas and 25 village panchayats occupies an area of 426 sq.km, that is 140% of its jurisdiction. The existing Chennai Metropolitan Area is 1189 sq.km as notified in 1975. The extend of Chennai Metropolitan Area is 1189 sq.km which comprises of one corporation, 16 municipalities, 20 town panchayats and 214 villages in ten panchayat unions. By considering the future development of the metropolitan area, the regional approach is adopted for an extending area of 8878 sq.km. The study area comprises of Chennai, Kancheepuram and Thiruvallur districts to analyse the metropolitan growth in a regional level. This study analyses the location advantages of different economic sources in relation to the resource. Based on the strategic location, this region further more serves the India's key economic generating cities namely, Bangalore, Hyderabad, Madurai, Vijayawada and Coimbatore, all lie within a radius of 400–500 k. Proximity and new opportunities offered with regard to the capital city will attract skilled and talented human resources from these cities nearby towns and villages. As being an important centre of employment, it attracts immigrants from all over the world, thus also improves the local economy.

2.1 *Existing Population Distribution*

Demographical study illustrates the changing structure and composition of a particular human population. As per census 2011, the population of the Chennai

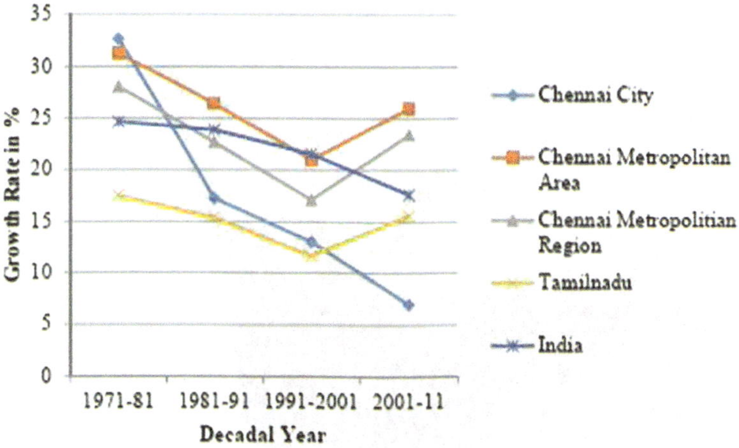

Fig. 1 Comparative tread analysis of interdecennial growth rate of population

Metropolitan Region(CMR) is 1.28 crore which contributes 17.7% of population in Tamil Nadu in which Chennai district contributes about 37% and considered as a highly populated area. Figure 1 shows the comparative trend analysis of interdecennial growth rate of population.

From the analysis, it is observed that both Chennai Metropolitan Region (CMR) and Chennai Metropolitan area (CMA) has a steady decline in the growth rate from 1971 to 2001. But in 2001–2011, CMR had a sudden increase in growth rate of nearly 6%. The Chennai city shows a drastic reduction in growth rate from 33 to 7% over a period of four decades. This shows that there is an increase in population in the rural areas in CMR as well as CMA which shows a same pattern of change.

Figure 2 shows the evolutional growth of built-up area within the CMA Limit and Tables 4.1, 4.2 and 4.3 indicates the conversion of farmlands over the decades. From Table 4.1, it indicates that growth of urban area increases from 35 to 52% in CMA. Once the growth of population gets stabilised in core area, people are showing interest to shift from core to peripheral areas in search of affordability. From Tables 4.2 and 4.3 denotes that conversion of built-up, agriculture, vegetation, water bodies and barren land. It is observed that growth and development are taken place along the transport corridor in the rapid manner. Other than the transport corridor, the development is taken place in scattered manner.

It results in fragmented urban growth in the outskirts of Chennai city with the transformation of green cover and agriculture land into built-up settlements. If current status of urban growth continues without proper direction, it will cause serious impact on CMA and it paves way for the low density and unplanned development. To mitigate urban sprawl and to avoid haphazard development, city intercellular approach is required to regulate the development in the planned

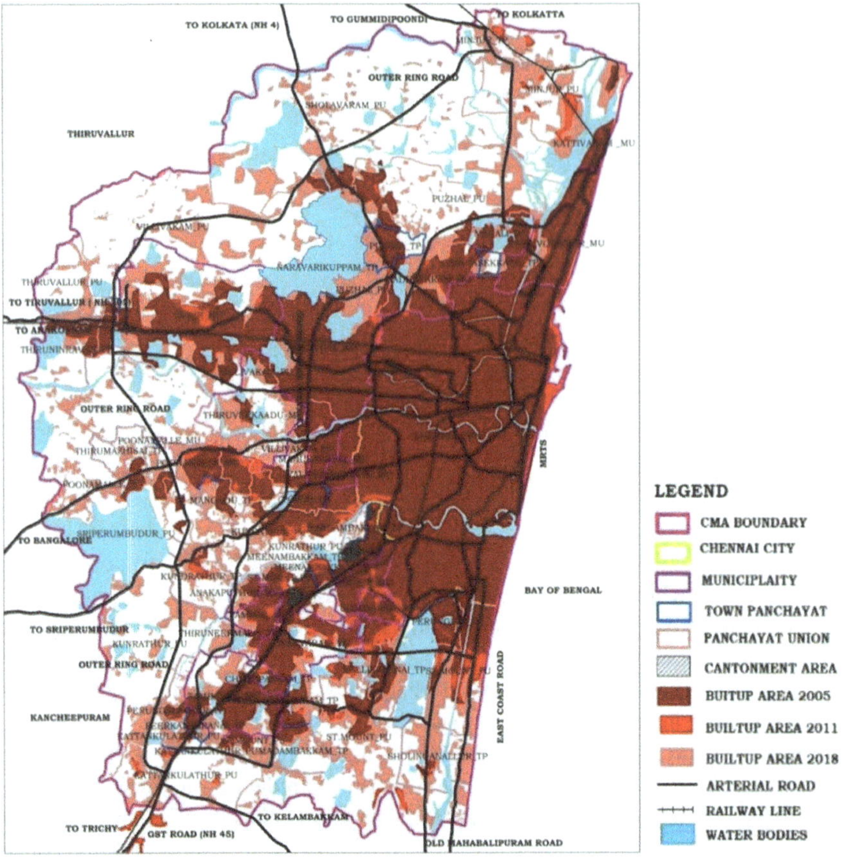

Fig. 2 Comparison of built-up area within CMA limit

Table 4.1 Built-Up area calculation

S.No	Year	Built-up area (sq.km)	Built-up area (%)
1	2005	410	34
2	2011	497	42
3	2018	618	52

Table 4.2 Landuse land cover in hectares

S.No	Description	1991	2003	2016	2027
1	Built-Up	20110	39965	58030	70836
2	Agriculture	9387	22857	5584	701
3	Vegetation	21667	9296	11754	3961
4	Water bodies	5320	6772	6120	5420
5	Bare land	25992	3596	998	1570

Reference: Modelling urban sprawl using remotely sensed data—a case study of CMA

Table 4.3 Land use land cover change in hectares (%)

S.No	Description	1991–2003	2003–2016	2016–2022
1	Built-Up	99	45	22
2	Agriculture	143	−76	−87
3	Vegetation	−57	26	−66
4	Water bodies	27	−10	−11
5	Bare land	−86	−72	57

manner. The complexity involves many aspects and components such as economic, social, physical and institutional. Complexity can be managed through collective intelligence to manage the difference.

3 Analysis of Urban Fabric

According to Victor (2012), a city is not just an evolved product of a day or a decade. It needs generations and generations to compose a city [1]. Settlements initially started as a clusters and development oriented along the transportation corridors, where the spatial jurisdictions are limited by certain distance. Urban morphology is the continuous process of land development which attains a specific spatial pattern based on the influencing factors. The chronology of Chennai's spatial development from Chennai Metropolitan Area (CMA) as the Central Business District (CBD) and the further developments emerged through the transit lines, thus results in outward spread of suburban areas. In between the railway lines, new arterial roadways were formed in a radial manner. Hence the entire metropolis developed as a radial pattern, which accelerates the city's expansion. The existing spatial distribution of Chennai city follows the radial model. Radial metropolitan pattern further extrudes congestion in the city core, as the development authority is unable to control the land development in the major nodes. It will finally end as an unsustainable mono centric pattern of development. Based on the hierarchical study of existing settlements using Cumulative Functional Index (CFI) technique, potential growth centres were concentrated within the CMA and the peripheral region remains undeveloped thus resulting in lopsided development of the region. The radial spatial development creates a congested development only along the corridors and on the other grade settlements remains in a dispersed un-served service area. The failure of radial pattern is due to supply being unrelated to demand. It doesn't respond to the needs, preferences, and potential demand of the Chennai Metropolitan Region. These untapped backward settlements may enhance with the help of hierarchical settlement orders, linking of transit networks and stipulating the economic activities through certain specific model.

4 Sustainable City Intercellular Approach

The metropolitan centre plays a significant role in generating income and production. The Gross Domestic Product (GDP) of any large Indian cities is as great as national GDP. Undoubtedly, the rural development is necessary for our progress. But it is no substitute for solving the urban problems. On the other hand, rural prosperity depends upon what happens to our cities, how well we organise and manage them. The cities afford the opportunity for those who come as well as those who are left behind. The dramatic increase in the size of metropolitan cities has not reduced the pressure of population or poverty in rural areas. Such problems not only of 'teaming cities' but also of 'teaming countrysides'. Sustainable intercellular approach deals with the efficient placement of land use activities, infrastructures and settlement growth over a large area than an individual parcel of land.

The main objective of this model focuses on the following aspects. Such as

- To adopt a settlement hierarchy in the Chennai Metropolitan Region for balanced development to spread the economic activities and civic amenities across the region to their hierarchy.
- To promote economic growth by focusing on certain key secondary and tertiary sectors.
- To provide safe, reliable and efficient multi-modal transportation network and ensure good connectivity to every part of the region.
- To suggest urban planning strategies for combating and mitigating undesirable consequences due to urban growth through intercellular model.

The city intercellular models mainly adopt two principal components such as spatial integration and level of influence. City model insists the alternate approach to Walter Christaller's Central Place Theory. But, this model adopts the basic strategic principals of Christaller's spatial theory. Further sustainable intercellular approach identifies the most appropriate strategic location for economic development. It vertebrates the system in a natural way and responds to the local needs. Thus it restraints the historic planning concepts of hexagonal, linear or circular approaches. Spatial development predominately follows the linear pattern due to the existence of transit corridors. City intercellular suggests to response such linearity should follow some gradient. Example, land proposed for future development should be parallel to the main feature of a transversal force gradient. The metropolis has a different scale of urban units; to amalgamate such units will be possible through the hierarchical ordering of the settlements.

Sustainable city intercellular approach has been described as the regional land use matrix. It defines the development along the transit corridors which generates spatial gradient parallel to the economic sectors and perpendicular to the remote settlements. These cells constitute an integration of urban features. These cells define the scale of urban form and urban fabric. The development of sustainable city has the ability to build up an economically efficient, socially equitable and

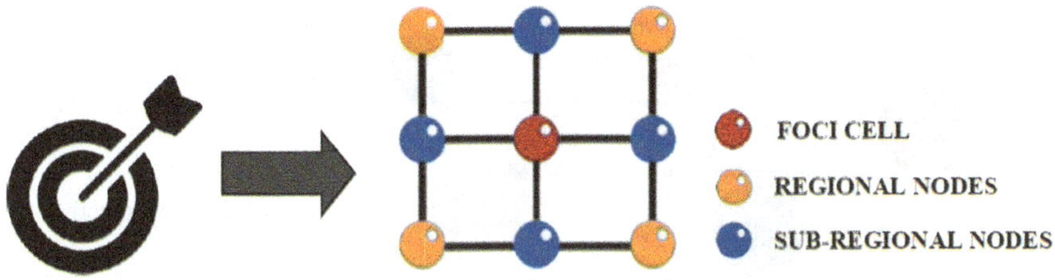

Fig. 3 Transformation of uni-focused planning to the cellular approach

environmental-friendly metropolis [2]. It has various hierarchy of cells and it is designated based on various sectors as follows, Central Governance District as a foci cell followed by regional and sub-regional nodes. Each node has its own focus of development criteria. Figure 3 shows the transformation of uni-focused planning approach to the cellular approach.

Foci Cells: These are metropolitan centres, acts as a Central Administrative/ Governance District, which is bounded by major transit networks and has higher order of facilities.
Regional Nodes: Regional nodes are the secondary level of cells, which has the potential to attract the economic development.
Sub-Regional Nodes: It serves the regional nodes and the average influence of this zone range from 20 to 14 kms.

The entire region and its environs shall be taken as a single planning and development entity. The location of each cell determines its own territorial control and their capacity to build the sustainable region. Spatial specific approach includes the zoning of land use and its related economic activities. This helps to avoid the mixing of incompatible spatial development. It further helps to promote higher-density development on and along the economic centres or over the regional nodes.

4.1 Significance of the City Model

Sustainable city intercellular model also suggests the location specification of these cells based on their urbanisation status. This framework also suggests the spatial equilibrium. Figure 4 shows the equilibrium status of the intercellular city model.

Cellular model controls the dynamic development and assumes that the equilibrium between each cell could rectify the supply and demand of the influencing area. It considers various scale of processes of urban changes and concentrated on their own. Though the cells may vary from its scale and its function, the diversified cell distribution with its economic generating and service aspects achieves the equilibria state.

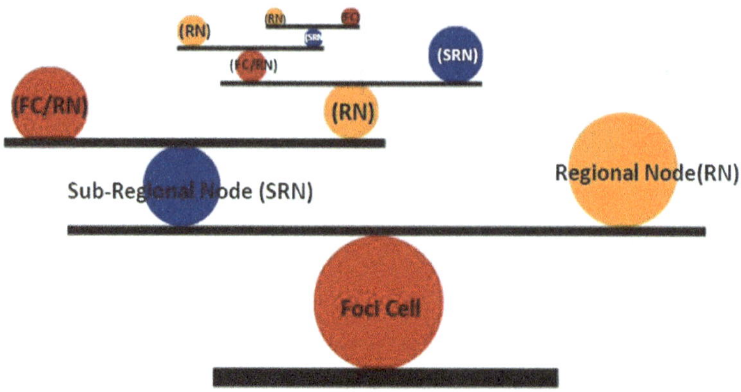

Fig. 4 Cellular equilibrium

The development phase of city model disparate the governance, spatial development and its associated users as shown in (Fig. 5). In the existing trend of planning approach, the governance and the spatial development do not correspond to the pattern of living and distribution of population. This model insists the integration of User behavioural pattern along with governance and land use development of a specific area. It also varies in scale from foci cell to the central villages.

The foci cell and the related adjoining regional centres are no longer distinct. They are interlinked by an intense system of relationship which is ever growing because of the recent phenomenon of urban growth. A strong community of economic and social interest joins the fringe and urban areas together. As the specialised services are provided by the core urban area, the cost and benefits of these services spill over the metropolitan boundaries.

In spatial term, cells are so because they benefit from location advantages. They have grown because of their strategic location controlling the interface between two ecosystems. This provides an intercellular structure, unlike the orbital one of featureless unadorned cities [3]. Conurbation has been controlled with the help of polycentric intercellular approach.

5 Formulation of Model

Sustainable city intercellular model built on stepwise approach is considered as the suitable form to analyse the spatial development involving a various number of variables.

- Identify the potential areas for spatial growth and development—parameters taken for analysis—growth direction, growth rate and population density.

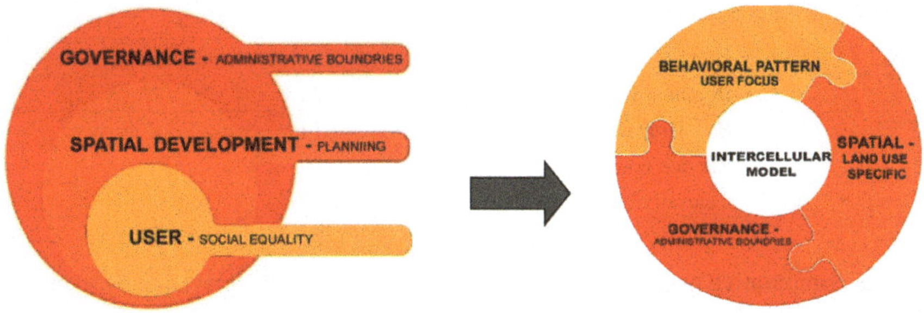

Fig. 5 City sustainability through intercellular

- The shape of the city is the result of the private sectors and of the modest government.
- A city's spatial pattern gives it form and beauty, and they determine the city's affordability and access to jobs.

Range of various parameters adopted for the study includes:

- Demographic factors
  - Population size
  - Growth rate
  - Density
- Economy factors
  - Work force participation
- Basic facilities
  - Presence of institutional facilities
  - Availability of electrical connections (industrial, commercial)
- Accessibility
  - Pucca road per unit area
  - Proximity to railway station

Ranking of settlements is essential to provide opportunities and facilities even to the least order settlement. Cumulative functional index technique has been adopted for determining the growth performance of centres.

The following parameters considered to assess the Hierarchy of the settlement are demographic, work force participation, presence of basic facilities and connectivity to the settlements. This technique is regarded as one of the most rationale method to identify the functional and growth performance of the settlements in a given region. Demographic indicator and service indicator are the key parameter to assess and identify the order of settlement.

Table 4.4 Influence radius of the settlements

S.No	Hierarchy of the settlement	Influence radius (Km)
1	Regional nodes	20
2	Sub-regional nodes	14
3	Service nodes	10
4	Central villages	5

The comparative assessment of performance index can be arrived at by using the following formula.

$$P.I = \frac{N - \text{Min}}{\text{Max} - \text{Min}} \times 100$$

where

N = Observed value of each indicator against each settlement
Min = Minimum value of the selected indicator
Max = Maximum value of the selected indicator
P.I = Performance Index

The calculated performance index value is summed up for different facilities as per the census data 2011 and the cumulative functional index (CFI) value is arrived, using mean and standard deviation to identify the range of hierarchical orders of the settlements. The zone of influence of each town of hierarchical order has been delineated using modified quantitative technique devised by V.L.S. Prakash Rao—delineating sphere of urban influence.

$$S.I = TC \times A/C$$

where

S.I.—Sphere of influence
TC—Centrality score of the regional nodes.
A—Total area of the region.
C—Total centrality score of all the nodes.

Cellular models suggest the up gradation of existing lower order services to higher order services to satisfy the spatial equity [4]. The sphere of influence of the settlements is found out as per the methodology and average influential radius of each order of the settlement is tabulated in Table 4.4.

5.1 Model Integration for the Study Area

Based on the analysis, one metro centre as Chennai Metropolitan Area, six regional nodes and 16 sub-regional nodes were proposed to achieve the spatial equality with an influence radius of 20 and 14 kms. Such arrived hierarchical order would be linked through the transit networks in an intercellular way. City intercellular model promotes industrial development on the major growth centres in which 80% consists of special economic zones. But the mono centric spatial development of industries results in lopsided development. Industries are nuclei of regional growth. In the intercellular, the major intersections are the regional nodes and the diagonals in between are those plasmatic transits. That promotes strategy and tactics for urban socio-economic development and growth. The concentration of clustered industries was more in Sriperambathur, Chengalpattu, Ambathur, Sholinganalur which help to strive the economy of the region due to the presence of special economic zones. The analysis results that more than 20% of employment generated from the regional nodes. These nodes agglomerate the economics in order to increase the productivity and reduces the cost of production and transportation cost. Such diversification offers the industry can leverage its core competencies and capabilities and offers networked employment which may be to cover the whole region and may able to achieve economic and spatial equity. Figure 6 shows the distribution of regional and sub-regional nodes and its influencing economical area for Chennai Metropolitan Region. The major intercells act as a transitional space and it varies in scale. Intersection on Metropolis through fares with commuter train potential services. The integration of peri-urban area can be done with the help of proposed nodes and the influencing areas is key to the high density residential and commercial development.

Land use zoning is the important tool in making the urban growth orderly and ensure that the metropolis does not decline in quality. Influencing area of two SR further enhance the green field development, whereas the higher order nodes promote the economical development.

6 Conclusion

The growth of population in the core city has led to the emergence of continuous urbanised zone and conurbation has a serious dimension of management and the provision of civic amenities and services in various cities. The urban scenario speaks about various facts of problems like poverty, traffic and transportation deficiency, escalating costs of developments and mounting prices of facilities which are fast growing beyond the reach of the several millions of the vulnerable section of the society. Hence, the urban development components are shelter, liveable environment, social and leisure facilities and facility for works like industrial and commerce development, etc. Intercellular model identifies the untapped potential area and propose various strategies to attain spatial equity for the Chennai Metropolitan Regional. Though the urban planning meet up with various city models, this intercellular gives the perspective of larger scale and also integrate various aspects to enhance the liveliness of the planning area. Focus is on the conservation of natural resources and allocation of resources at taluk level as only

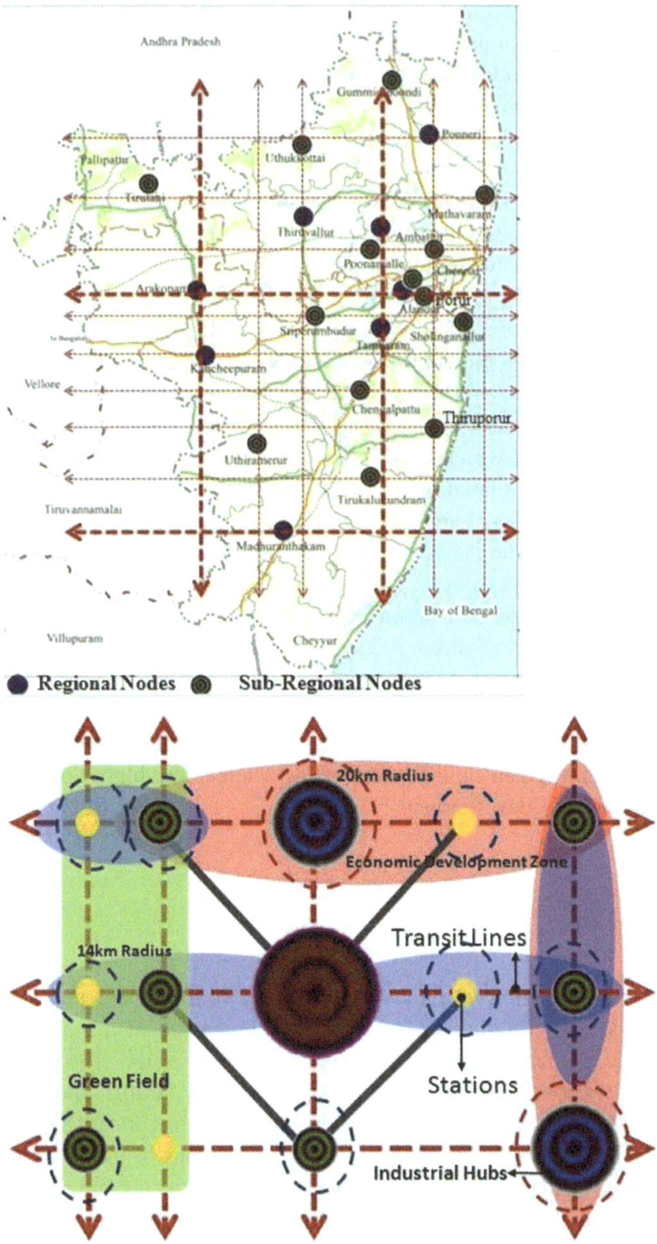

Fig. 6 Sustainable city intercellular model—Chennai

an aggregated level. The fundamental goal of this regional planning model is to rebuild the economy, spatial equity and environment inclusive development considering its unique natural and manmade features to promote both rural and urban areas. This study also identifies the strategic locations of the metropolitan development for a projected population.

References

1. Karthigeyan D (2016) Transformation of Chennai city as nucleus of regional development through the emergence of sub-CBD's. Int J Earth Sci Eng 14–20
2. A Global Report on Human Settlement. (2009) Planning sustainable cities, United Nations Human Settlements programme
3. Contin A, Ortiz Castagno P, Musetta A, Bovio S, Zammataro A, Frigerio A (2014) Cabodelgado 2015: growingsmart. Proposal for a scientific research and cooperation project for Cabo Delgado—Mozambique, I Conferenzaannualecentro PVS—Planning Viable Sustainability with Emerging Regions, La Sapienza, Rom
4. Ortiz Pedro B (2014) Metro matrix model for a non-fragmented city: the new metropolitan morph-type, such as a grey and green geographical skin of the infrastructure. EURAU, IIstanbul-Turkey

Sustainable Water Management: Smart Solutions for Equity in Vellore Municipal Corporation

Sugato Dutt and Prabhakaran Punniakotty

Abstract Tamil Nadu state envisages 24 × 7 piped water supply to all households in urban and rural areas in tune with SDG Goal 6 (established by the United Nations General Assembly in 2015) ensuring equity in supply of drinking water and sanitation. However, problems of sustainable water management have increased due to growing concentration of the state's population in urban areas like Vellore. Planning for sustainable use of water resources has gained significance due to agricultural intensification and industrial growth despite constraints in ground water aquifers. Large investments in supply side management for drinking water include feeder pipes from distant sources and intermediate sumps. Nevertheless, investments in demand side management including recycling and reuse remain unaddressed in planning for enhancement of per capita water availability. This paper provides a case study of Smart City initiatives in Vellore Municipal Corporation which has expanded to include six panchayats and two municipalities and is dependent on subsoil water sources in the absence of a perennial water course. It highlights the need to incorporate demand side management through inter-alia investments in water harvesting, storage, use efficiency, and effluent and sewage treatment in order to address equity concerns. It also discusses gaps in the present system while advocating sustainable withdrawal and supply of freshwater under seasonal scarcity conditions.

Keywords Sustainable water management · Smart city initiatives · Vellore municipal corporation · Smart solutions · IoT

S. Dutt (✉)
TNCAMPA, PCCF, Chennai 600015, Tamil Nadu, India
e-mail: dutt.sugato@gmail.com

S. Dutt
APCCF (Social Forestry & Extension), PCCF, Chennai 600015, Tamil Nadu, India

P. Punniakotty
Principal Architect, FSO Designs Pvt Ltd, Chennai, Tamil Nadu, India
e-mail: Karansmailbag@gmail.com

K. Thirumaran et al. (eds.), *Sustainable Urban Architecture*,
Lecture Notes in Civil Engineering 114,
https://doi.org/10.1007/978-981-15-9585-1_5

1 Introduction

Urbanization is indeed a characteristic of economic development. India's rapid GDP growth in the past few decades has resulted in a high proportion of urban population [6]. This is expected to cross 590 million by 2030. The processes of natural growth in population as well as increased rural to urban migration have resulted in the expansion of urban limits across the country [3]. A conglomeration of municipal corporations with adjoining census towns has become a characteristic feature of urbanization in the country today. Population growth, together with rising incomes in the resultant urban sprawl, has compelled increased consumption of natural resources such as minerals, forests, and water [10].

Today, surface water sources are being saturated and ground water is being constantly depleted and adequate mechanisms for its recharge place a considerable burden on the state exchequer. In addition as cities agglomerate, water from rural areas is being sourced to provide for the growing needs of the main urban centers [8].

The United Nations in its declaration of Sustainable Development Goals (SDGs) envisions three dimensions of sustainability, i.e., social, environmental, and economic, thus becoming crucial for the human development agenda. The SDGs are a successor to the Millennium Development Goals (MDGs) that have aimed to raise the human development index across the nations of the world. The 17 SDGs endeavor to add sustainability to the development agenda and allow a level of development that is fair to different sections of society (equitable) and economically viable in its utilization of resources. In the ultimate analysis, this is also a projection for environmental sustainability (Fig. 1).

Water, as a resource, lies at the intersection of several Sustainable Development Goals. Goal 6 deals with clean water and sanitation, Goal 14 deals with life under water, while Goal 1 itself deals with the elimination of poverty. Safe drinking water

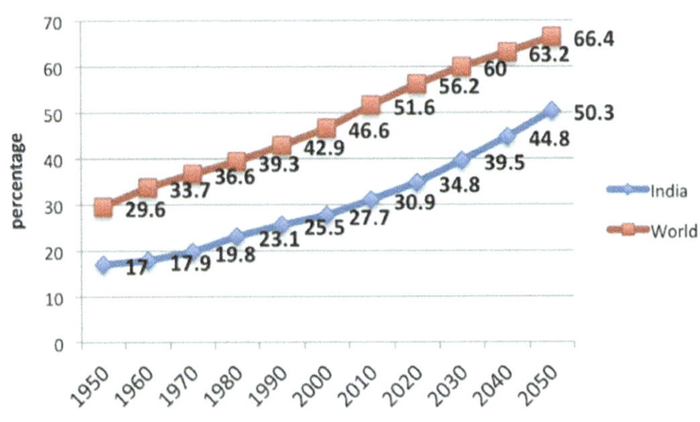

Fig. 1 Urbanization in cities from 1950–2050 *Source* OWID based on UN World Urbanization Prospects 2018 and historical sources

and sanitation for all is therefore a desirable end in itself but such water ensures health of all as stated in Goal 3. The equitable distribution of water and the eventual efficiency of its utilization help in health and well being, preventing disease and contributing to the reduction of poverty, while life under water in the form of fish stock as well as vegetation in the oceans, rivers, and other water bodies contributes to general food security. Overall assurance for water use efficiency by all sectors then ensures the sustenance of large sections of society, both urban and rural, vide Goal 11 (Fig. 2).

As such most urban local bodies in India are saddled with financial management problems. The high costs of water supply infrastructure, its maintenance, the politics of levying user charges and fees, and the difficulties in preventing leakages and diversions make the management of water supply a challenge to sustainable resource management for overall economic development (Fig. 3).

Water leakages, inequalities in distribution, and poor quality control are issues that plague water distribution networks in most developing nations. In particular, municipalities are hard pressed to ensure a steady supply of safe potable water. In addition, the difficulty of regulating the growth of urban agglomeration, the overall poor quality of meters, and the political challenge of acceptable levels of municipal tax are hampering the water use efficiency and sustainability of water management practices (Fig. 4).

The Govt. of India has launched the Smart City Mission in 100 notified cities in order to make them citizen friendly and sustainable. Several cities have adopted the development of infrastructure for water and sanitation (aside from other areas such as urban mobility, housing, energy management as also water management). The eventual purpose of the Smart City initiative is to drive economic growth and improve the quality of life of the people by enabling local area development by harnessing technology, i.e., technology that leads to smart outcomes. Robust IT connectivity and digitalization are to be the bulwark of smart cities with smart solutions for instance for water management being provided by smart meters for data acquisition and integration, cloud technologies for data processing and storage, and finally analytics and decision support using other web-based technologies.

Fig. 2 Sustainable development goals

Fig. 3 Water leakage

Fig. 4 Smart city mission

According to the Govt. of India's Technical Report on Designing and Planning Smart City Initiatives with IoT/IT, "Urban Design and Planning and Futures Thinking Principles coupled with Digital Communications and Technology Adoption is what will separate cities that are trying to achieve smartness from those that are truly smart, i.e., by harnessing technology to enable sustainable growth." Also, "smartness" is embedded in cities by capturing relevant data and understanding and analyzing behavioral and analytical trends [4].

This paper discusses the scope for using Smart City initiatives, in particular the Internet of Things (IoT) in dealing with the opportunity and gaps in the water sector with Vellore as a case study and it also provides a case study of Smart City initiatives in Vellore Municipal Corporation which has expanded to include six panchayats and two municipalities and is dependent on sub-soil water sources in the absence of a perennial water course. It highlights the need to incorporate demand side management through inter alia investments in water harvesting, storage, use efficiency, and effluent and sewage treatment in order to address equity concerns. It also discusses gaps in the present system while advocating sustainable withdrawal and supply of freshwater under seasonal scarcity conditions.

2 Literature Review

2.1 Areas of Intervention

The threats of water scarcity, aging infrastructure, and rising production costs are driving utilities to consider smart water solutions. According to the United Nations, global water use has been growing at more than twice the rate of population growth in the past century [2].

Gupta et al. [5] have identified key areas for smart water management such as control measures for loss of non-revenue water (NRW) through leakage detection; smart farming through sensor controls that reduce wasteful water use; sewage treatment and effluent treatment, etc.

2.2 Opportunities and Gaps

Future thinking and trend spotting: Smart solutions provide mechanisms to consider trends technologies and user behaviors. Individuals are known to frame problems according to their perceptions biases, filtering information, and distorting inferences [4]. Futures studies use "trend spotting" to detect these biases and provide solutions that are financially viable and environmentally feasible in the long run.

Using digital technology, a water smart city can sustainably use and reuse water resources by adapting real-time operations and planning practices in response to ubiquitous sensor networks and disparate but interconnected and heterogeneous data streams. While a survey of the water industry shows that 33% of utilities are interested in real-time control and big data system analytics, water utilities have predominantly not harnessed these technologies, due to a number of challenges associated with managing and analyzing big data. Technological gaps, workforce challenges, and community disengagement undermine the alignment of critical municipal management priorities with the analysis and application of smart water data.

2.3 Smart Solutions for Water Provisioning

A smart water network depends on how well a municipality manages its distribution networks, creates awareness regarding efficient use, provides safe water, manages leak loss, and generates revenue. Citizens themselves must learn to alter their behavior in order to reduce, reuse, and recycle water. Improved water quality, its access, and eventually increased use efficiency are key outcomes of such a smart strategy.

Inequality of water distribution: Municipalities in India are generally unable to provision water adequately for their consumers for various reasons including poor condition of pipeline infrastructure, inadequate revenues for their improvements Fear of resorting to unpopular means such as enhancing prevailing taxation rates has compounded the problem.

2.4 Basic IoT Model

Focus Group on Smart Sustainable Cities formed under International Telecommunication Union is carrying out a study since 2013 exploring smart water management for smart sustainable cities. The study suggests to include following components to include within the umbrella of smart water supply system:

1. Smart pipes and sensor networks
2. Smart metering (LAN-HAN-WAN) having pre-paid/post-paid arrangements for end user
3. Communication modems
4. GIS mapping and relational database
5. Cloud computing (analytics)
6. Supervisory control and data management (SCADA) models, optimization, and decision support
7. Web-based communication and information system tools.

A basic three-tier structure has been developed for the IoT related to smart water management [9]. The "bottom" tier consists of a set of objects or "things" including sensors, mobile phone actuators, and RFID systems. The middle tier consists of a network for the transformation of information that has been collected by this set of objects. The top tier is where analysis of information gathered is conveyed to the end user (Fig. 5).

The architectural model for smart water management incorporates a "physical" layer with the necessary components for delivering water (e.g., pipes, pumps, valves, pressure reducing valves (PRVs), reservoirs and other delivery endpoints) and a "sensing and control" layer with equipment and sensors that measure parameters (e.g., flow, pressure, water quality, reservoir levels, water temperature, acoustic information, etc.). This data is then transmitted and stored through the "collection and communications" layer which includes fixed cable networks, radio, cellular, and Wi-Fi (Fig. 6).

An inbuilt GPRS module enables connection to the cloud over the Internet. The equipment is battery powered, which can be automatically charged by solar power or AC supply. Data acquired from the sensor module can be calibrated against set standards World Health Organization, etc. Sensor modules are cloud-connected, and thus, any anomaly detected is notified to the municipality personnel and the end user via an app. If the anomaly parameter exceeds the threshold value, the supply

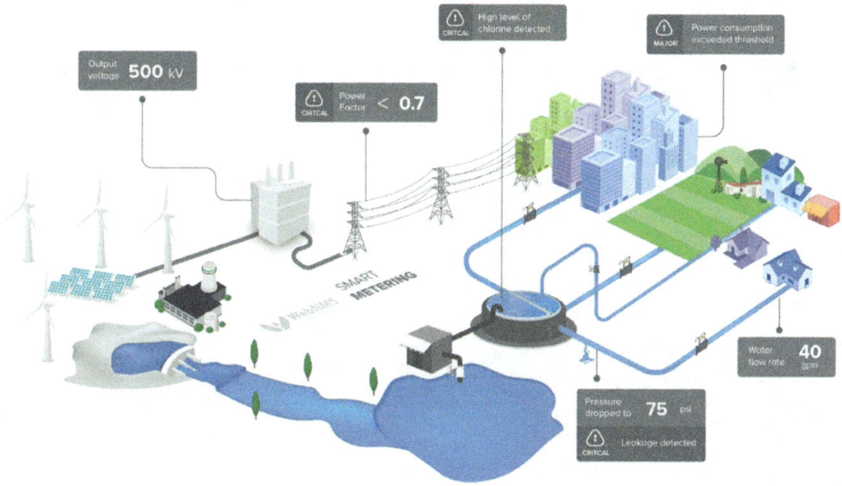

Fig. 5 Smart metering

inlet valve is being turned off automatically in case of end branches of the water supply system.

The sensor module can measure turbidity, pH, TDS, pressure. More parameters, like the presence of chemical contaminants, e-Coli, and microplastics, could also be detected by adding appropriate sensors. The modules, fitted at various nodes in the water distribution system beginning at the outlet of the treatment plant until the outlet at the end user, improve the coverage of water quality monitoring considerably over the existing ones. This system helps in conducting a preliminary test of water quality.

The fourth layer, "data management and display," aggregates data from the above three layers to create an interface with human operators, e.g., supervisory control and data acquisition (SCADA) system, GIS, network visualization tools, and water balance applications.

The final layer, "data fusion and analysis," is where more sophisticated processing of raw data occurs. This may include, for example, real-time data analytics, hydraulic modeling, network infrastructure monitoring, or automatic pressure, and energy optimization systems.

3 K-Water Case—Seosan City, South Korea

The Smart Seosan City project was established in January 2016 when the municipality of Seosan city asked K-water (who operated the local waterworks system in Seosan city) for a smart metering system for the local water supply system as a drought measure. Since 2015, the Boryeong Dam, which supplies water to Seosan

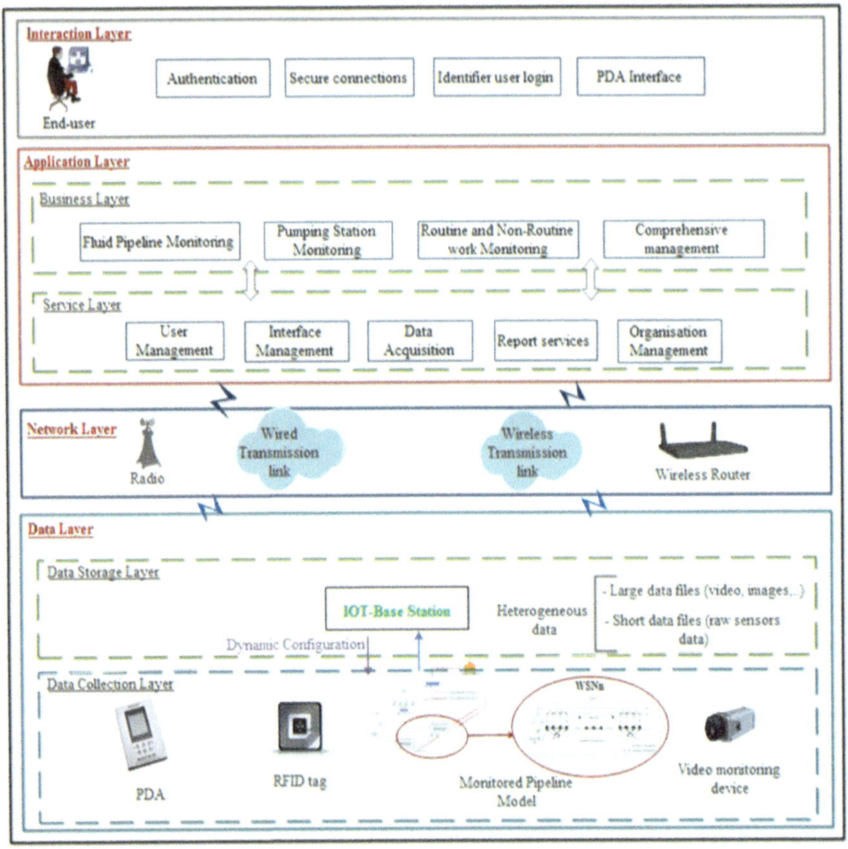

Fig. 6 IoT model

city, had reached a water storage minimum of 21%, resulting in the need for urgent measures to be taken. Seosan city invested approximately 500 million won in this project (equivalent to approximately USD 467,500), for K-water to integrate ICT technologies such as smart metering, wireless data transmission and decision making system to reduce water leakage rate of pipelines as a way to secure water in Seosan city.

Innovative smart water management technology solution proposed: To solve the various water-related problems and improve the efficiency of water management in Seosan, K-water considered various technologies. Recently, information and communication technology (ICT) had been adopted to maximize efficiency in the process of water production and distribution [1]. In Korea, there are many projects working together to develop a Smart Water Grid (SWG) (Byeon et al. 2015) and technologies that enable sustainable water supply by connecting water sources and optimizing water treatment based on ICT support this effort.

Water management using ICT, also known as smart water management (SWM), enables sustainable water supply to every citizen by water resource monitoring, problem diagnosis, efficiency improvement and harmonizing management, resolving many water related challenges [7].

Outputs and outcomes were decrease in leakage: After applying smart metering and the SDMA system to the target area, the revenue water ratio reached 90%. This was an increase of 20% compared to the first half of 2016 and of 19% compared to that of the previous year.

Inner leakage detection of users and restoration: Time-based use pattern analysis of the smart meter monitoring system enabled interior leakage for the user to be found and restored. As a result, each user saved 55% of water use and 70% of water cost.

Increase in customer satisfaction: By switching to using remote meters for water use and quality, customers' satisfaction has been improved as it has become possible to handle complaints promptly with surveyors responding to every customer concern and also by providing additional water quality management services such as the inspection of customers' indoor pipes.

In addition to this, as the target area for this project was based in a hillside area where the households are quite far apart it was originally too costly for the civil service to visit each traditional meter themselves to monitor the water use. This meant that customers were required to monitor the meters themselves taking them a lot of time and effort. By introducing remote metering, it enabled improved monitoring of the meters by the civil service, and as a result, the customer satisfaction has increased (Fig. 7).

4 About Vellore

Vellore is geographically located at 12.92° N 79.13° E, 220 m above the mean sea level. It is located about 145 km west of the state capital Chennai and about 211 km east of Bengaluru. The city lies in the Eastern Ghats region and Palar river basin.

The topography is almost plain with slopes from west to east. Vellore has four zones with totally 60 wards which covers an area of 87.73 km^2 and has a population of 5.04 lakhs as per census 2011. The recent population growth of Vellore City is shown below (Fig. 8).

As of 2001, out of the total area, 69.88% of the land was developed and 31.12% of the city remained undeveloped. Out of the developed area, 55.76% was used for residential purposes, 8.34% for commercial, 1.58% for industrial, 3.3% for educational, 16.46% for public and semi-public, and 10.12% for transport and communication. The population density is not uniform: It is high in areas like Arugandhampoondi and lower in the peripheral areas such as Poonthottam.

City Administration: Vellore is the headquarters of the Vellore District and promoted to a municipal corporation from August 1, 2008.

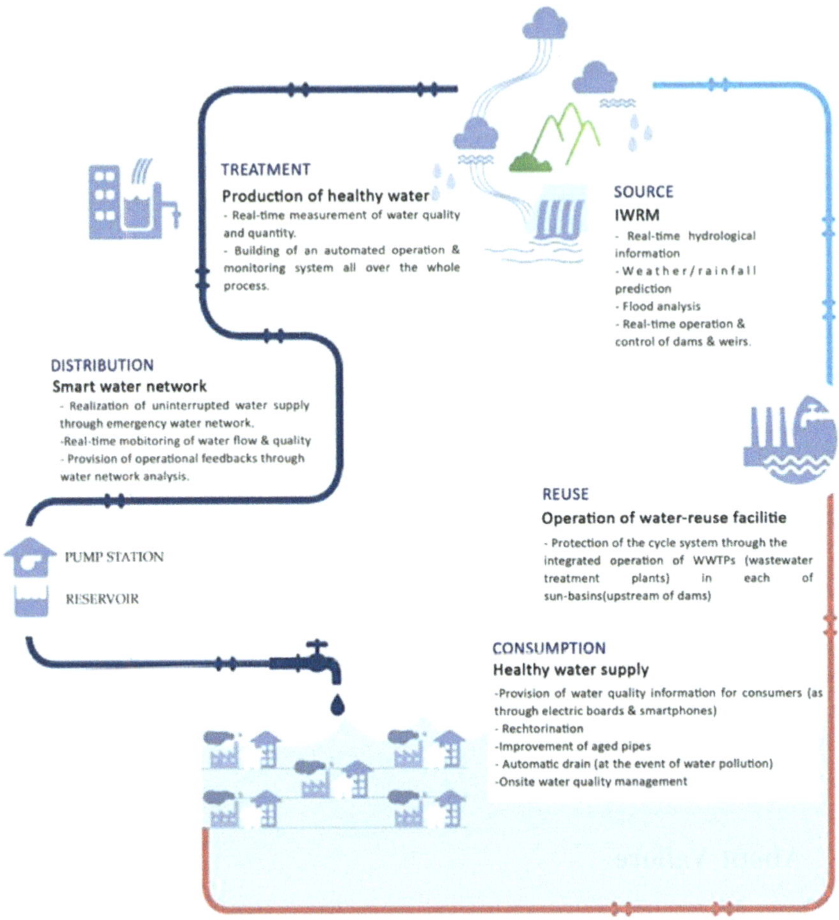

Fig. 7 SWM concept map

4.1 Link Between Vellore Municipal Master Plan and Developmental Plans to Water Sources

The Vellore Corporation intends to improve its water supply distribution network equity, in order to ensure daily availability of protected water supply to all. Further, the existing distribution network is having uneven water supply to users particularly on tail ends. There are certain areas where water supply is more than adequate in the system. Further, the corporation is fed with inequitable supply of water and poor demand coverage. Hence, it is necessary to improvise the existing water supply system and also develop a water supply improvement scheme for the distribution network. From sustainable water management and to improve the existing water

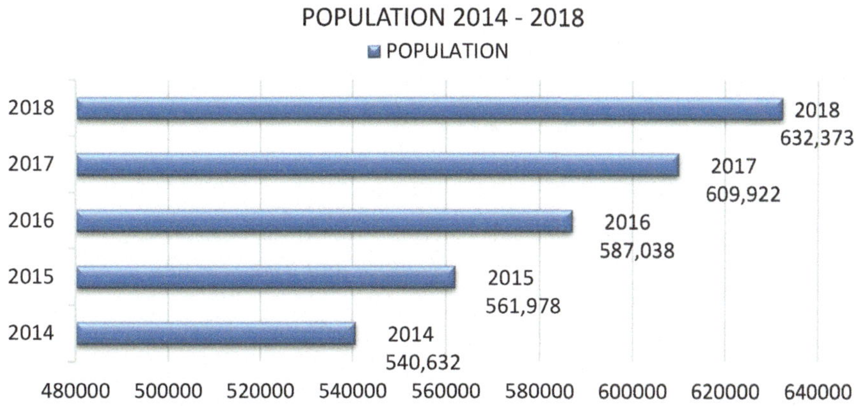

Fig. 8 Population of Vellore from 2014–2018

supply in the corporation area to achieve equi-distribution of water supply to all areas (Fig. 9).

4.2 *Necessity of Sustainable Water Management*

The water demand for entire corporation area was now assessed based on the following norms as given in CPHEEO Manual:Base water demand—135 lpcd, Fire demand—100 $(P/1000)^{0.5}$, Distribution losses—10%, Transmission Losses—1%, Water Treatment losses—3% (Table 1).

Presently, there is not any gap up to the intermediate year 2031, but there is a major difference when projected for 2046.

Water Leakage

Water users including industrial, commercial, institutional, and residential in Vellore are currently charged on a flat rate. The corporation supplies 114 LPCD, 30% of the water distributed is NRW. Water supply augmentation initiative worth Rs 252 crore water supply to 68 MLD and rehabilitate and upgrade the network infrastructure is underway. Source: Atal Mission for Urban Rejuvenation (AMRUT) SLIP, Vellore, November 2015. Water contamination is high in Vellore district and also in Vellore city as per TWAD (Figs. 10 and 11).

These are the drawbacks for the Vellore city. So, Vellore needs sustainable water management system to solve the problem in the city.

According to the Tamil Nadu Water Supply and Drainage Board (TWAD), the water supply distribution network now functioning in Vellore Old Municipal areas is more than 100 years old. The old and worn out distribution pipes are unable to withstand distribution pressures of the present age. Vellore Corporation meets its

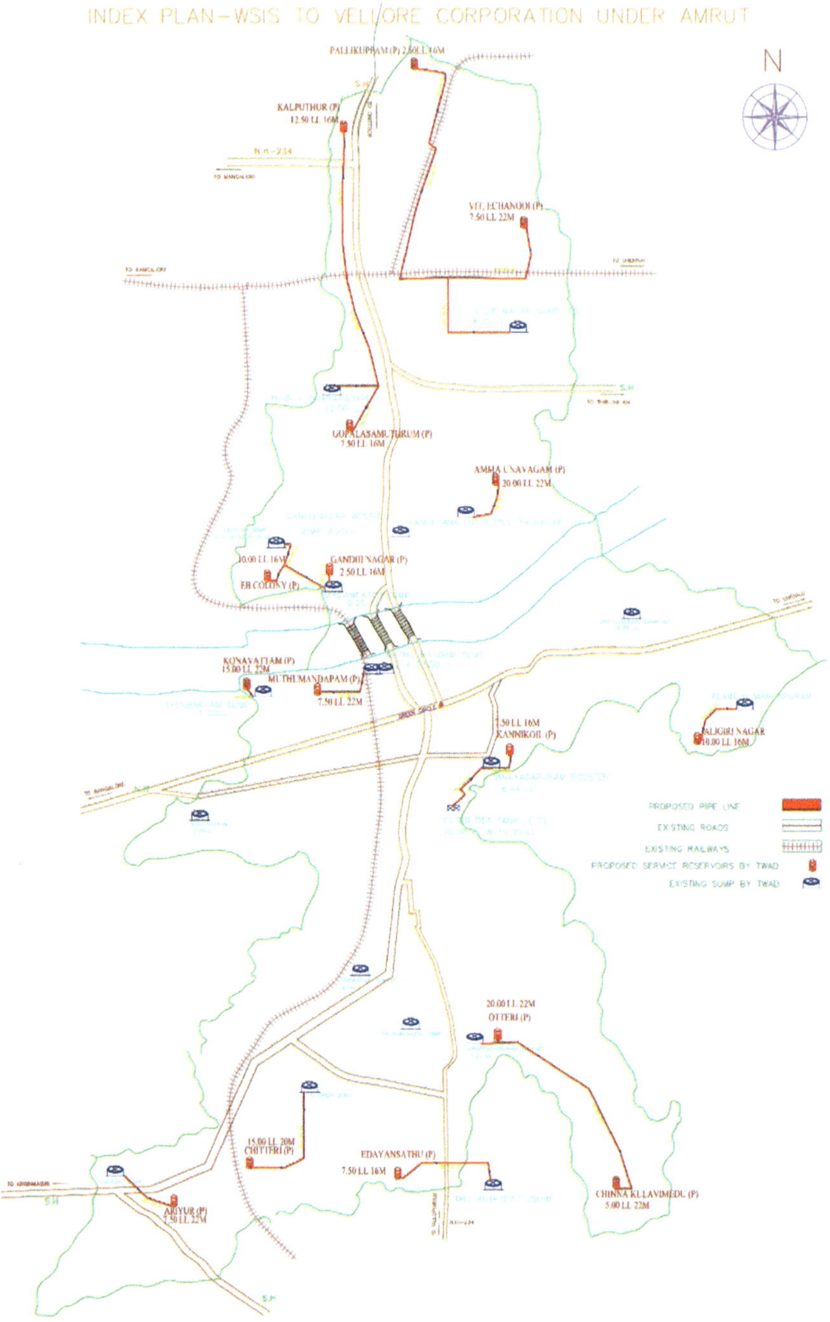

Fig. 9 Index plan—Vellore water supply system under AMRUT scheme

Table 1 Water demand, available source and deficit

S. No.	Year/Stage	Projected population	Water demand at 135 lpcd (Mld)	Available source (Existing + Proposed) (Mld)	Deficit (Mld)
1	2016 (Base)	5,49,349	78.647	86.43(27.43 + 59.00)	+7.783
2	2031 (Intermediate)	7,19,957	102.673	104.43(27.43 + 77.00)	+1.757
3	2046 (Ultimate)	9,20,000	133.363	104.43(27.43 + 77.00)	−28.933

Fig. 10 Water leakage in Vellore

water supply requirements through sub-surface sources, which include bore well/ open well at the Otterilake, Palar river bed, Karugampathur head works, and Ponnai river head works. The town is disadvantaged since it does not have perennial water course. The Palar river passes within the boundary of town but it is almost dry throughout the year.

Not only is the corporation fed with inequitable supply of water, there is also a poor demand coverage. Losses through leakages, inadequate water metering, and inadequate tax realization particularly in expanding and yet unaccounted new areas under the corporation are adding to the problems of water management. In addition, surface and ground water pollution from leather tanneries in Vellore continue to pose a challenge despite much litigation and also interventions to alleviate the problem.

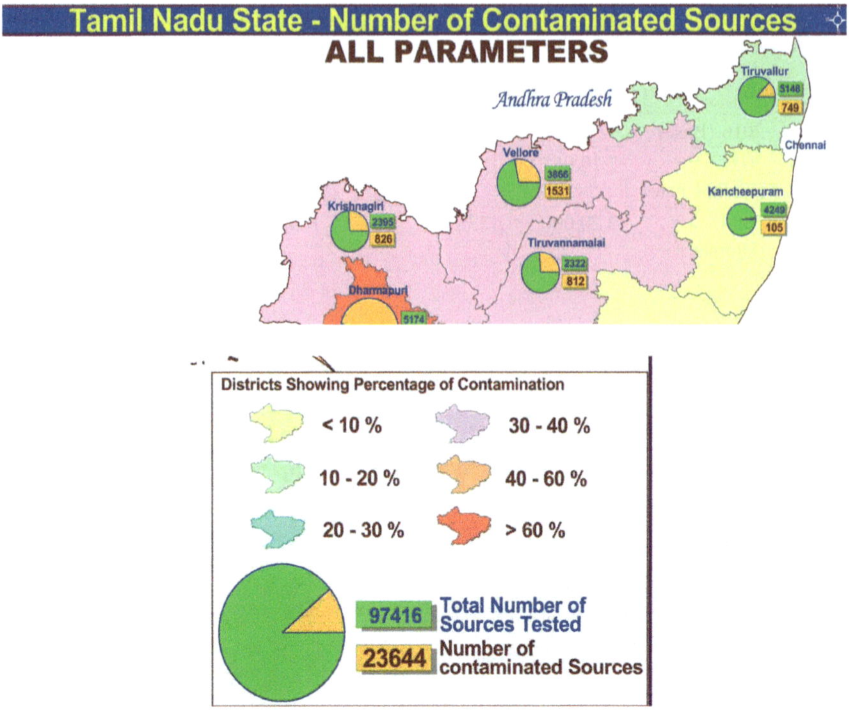

Fig. 11 Water contaminated sources

With a view to improve the existing water supply systems, the Tamil Nadu Water Supply and Drainage (TWAD) Board has projected the gap between water supply and demand well into the next few decades. The need remains to map and analyze the feeder main and distribution systems, reduce losses in transmission, eliminate pollution, and recharge the sub-surface storage systems and sumps through appropriate watershed and river basin management. Moreover, as the urban limits expand, there will be a high per capita demand for water and a greater need to address the distribution inequity.

5 Conclusion: Scope of Smart Water Initiatives

Water use efficiency through IoT has begun to be attempted by the Vellore Municipal Administration. Smart water metering has been initiated as elsewhere in Tamil Nadu. Similarly, SCADA or the supervisory control and data acquisition system to continuously monitor the flow of data on water flows, performance and efficiency of water pumps and motors, physical and chemical quality parameters in the water supply schemes of Tamil Nadu has been established. Likewise, water use efficiency measures in farming are taken up through initiatives such as system of rice intensification, drip irrigation, fertigation, etc. In case of sewage and septage management, a significant quantum of groundwater pollution may also be attained. Nevertheless, the measures outlined above are necessary to regulate and manage the demand for water in the district.

The incorporation of IoT approach in water management needs not to become a "panacea of all ills." Nevertheless, seeing the critical water availability per capita in Tamil Nadu, it is evident that water use efficiency should be enhanced by a systematic plan that integrates varied water using agencies and addresses the gap between demand and supply in a meaningful and reassuring manner. In this scenario, it is observed that the Smart City Mission has identified a number of sectors in urban planning including transport, electricity, solid waste management, energy use in addition to retrofitting and redevelopment of existing urban infrastructure. The agenda of water management particularly that of addressing the gap between demand and supply is relatively less strongly emphasized.

Smart city of Vellore provides water supply projects which are stand alone in nature (providing smart meters, upgrading infrastructures etc.). However there is less spending on integrating the whole system. IoT fills the gap to have a sustainable solution.

Thus, looking into the situation pertaining to urban Tamil Nadu, an incorporation of the above is well warranted. A smart city sub-mission dealing exclusively with water use management including its reuse and recycling is the need of the day. The Chinese model of Sponge City Initiative, which incorporates inter alia urban planning and smart water management is set of address the problem of urbanization impacting water resources may well be adopted for this purpose [11].

Acknowledgements Author is indebted to Ar. Mohammed Bayaz B. Arch Junior Architect, FSO Designs Pvt. Ltd, and S. Jeya Chandran Student, MIDAS School of Architecture for assistance.

References

1. Byeon S, Choi G, Maeng S, Gourbesville P (2015). Sustainable water distribution strategy with smart water grid. Sustain 7(4):4240–4259
2. Cahn Amir (2014) An overview of smart water networks. J Am Water Works Assoc 106:68–74. https://doi.org/10.5942/jawwa.2014.106.0096
3. Census (2011) Office of the registrar general & census commissioner, India, 2011 Census Data, Population Enumeration Data
4. Government of India (2019) Technical report on design and planning for smart cities with IoT. Department of Telecommunications M2M smart city working group
5. Gupta A et al (2016) Need of smart water systems in India. Int J Appl Eng Res 11(4): 2216–2223
6. Haque I, Patel P (2018) Growth of metro cities in India: trends, patterns and determinants. Urban Res Pract 11(4):338–377. https://doi.org/10.1080/17535069.2017.1344727

7. Heland F, Bondesson A, Nyberg M, Westerberg P (2015) The citizen field engineer: crowdsourced maintenance of connected water infrastructure. In: Proceedings of 29th International Conference on Informatics for Environmental Protection and the 3rd International Conference ICT for Sustainability, Atlantis Press, Copenhagen, 7–9 September 2015, pp 146–155
8. Rodrigo G (2004) Water, water, nowhere: a case study of Palayaseevaram village regarding sharing of water with the Chennai city and its impact on the village. In: Forum for Common Resources Concern (FCRC), a program unit of Gandhian Unit for Integrated Development Education (GUIDE), Chengalpattu, Nadu, India. WHIRL Project Working Paper, vol 9
9. Shahanas KM, Sivakumar PB (2016) Framework for a smart water management system in the context of smart city initiatives in India. Procedia Comput Sci 92:142–147
10. Sridharan N (2011) Spatial Inequality and the politics of urban expansion. Environ Urbanization Asia 2(2):187–204. https://doi.org/10.1177/097542531100200204
11. Zevenbergen C, Fu D, Pathirana A (2018) Transitioning to sponge cities: challenges and opportunities to address urban water problems in China (2018):1230

Redefining the Relationship Between Heritage and Its Community for Sustainable Development: Taking Temples as Case Example

Meera Viswanath and Nishant

Abstract Conservation in a broad sense refers to the protection of resources which are important for the future generation; let it be natural resources, energy, or heritage. When it comes to Heritage Conservation, the term Heritage has been defined and redefined by various organizations to broaden its scope; first, it was defined as Historic Monument in 1964 (Venice charter), got reinterpreted by ICOMOS in 1965 as Monument and Site, and again, by UNESCO in 1968 as "Cultural Property." The word cultural property implies its geographical setting as well as the people around, by which it recognizes the fact that every heritage shares an inseparable relationship with their immediate neighborhood, which in its true sense forms the soul of the chosen property. In modern India, identification and protection of these cultural properties have been carried out since the colonial times, in which the property is being viewed as only a monument. Indeed, its protection and continuity of the property at a skin level is important and must be ensured, but it should also be made sure that the process of protection does not uproot the property from its immediate context. Putting up a protective fence all around the monument for the sake of protection (a typical measure taken up by the government bodies in India) not only creates a physical barrier between the building and its context, but also alienates the people living around since ages; the direct/indirect association of the people with the monument gets broken and then dies gradually, leading to a complete disconnect. This paper focuses on critically analyzing this relationship, which is an essential understanding required for developing a holistic/integrated heritage management plan for cultural properties, which in turn will lead to a sustainable development. Temples are one of the best examples in India, which explains and exhibits the level of interdependency that exists between a cultural property and its people. Taking Indian temple heritage as case example, the paper tries to understand the idea of Heritage in Indian context and the association these

M. Viswanath · Nishant
National Institute of Technology, Calicut, India
e-mail: meeraviswanath@gmail.com

Nishant (✉)
R.V. College of Architecture, Bangalore, India
e-mail: nishant15bhardwaj@gmail.com

K. Thirumaran et al. (eds.), *Sustainable Urban Architecture*,
Lecture Notes in Civil Engineering 114,
https://doi.org/10.1007/978-981-15-9585-1_6

cultural properties have with the context. It also examines the disconnect caused between these cultural properties and their context by means of insensitive interventions, and the economic disparity created within the context as a result of the myopic policy/proposals created for development and management of these cultural properties.

Keywords Heritage conservation · Temple heritage · Integrated heritage management

1 Heritage and Conservation

1.1 Understanding the Term Heritage

The term heritage is so deep and vast that it cannot be defined with a single definition; several national and international organizations have tried to define and redefine it from time to time. It was defined first as Historic Monument in 1964 in the Venice charter, got reinterpreted by ICOMOS in 1965 as Monument and Site, and again, by UNESCO in 1968 as "Cultural Property." The word cultural property implies its geographical setting as well as the people around, by which it recognizes the fact that every heritage shares an inseparable relationship with their immediate neighborhood, which in its true sense forms the soul of the chosen property. In simple terms, it is the tangible (can be perceived physically) and the intangible (cannot be perceived physically, but which is experiential) aspects of an individual as well as collective memory.

1.2 Why Should We Conserve Our Heritage?

Dr. APJ Abdul Kalam once said, "We will be remembered only if we give to our younger generation a prosperous and safe India, resulting out of economic prosperity coupled with civilizational heritage."

The term conservation has been defined by UNESCO as "Measures taken to extend the life of cultural heritage while strengthening transmission of its significant heritage messages and values." In the domain of cultural property, the aim of conservation is to maintain the physical and cultural characteristics of the object to ensure that its value is not diminished and that it will outlive our limited life span.

A historic building to common knowledge is one that gives us a sense of wonder and makes us want to know more about the people and culture that produced it. It has architectural, aesthetic, historic, documentary, archaeological, economic, social, and even political and spiritual or symbolic values, but the first impact is always emotional. These are the attributes of our memory of our past, and it is the duty of each generation to hand down its heritage to the next.

Historic buildings hold a town together and establish a continuity between the past, present and future at a time where continuity is a scarce commodity. Maintaining a sense of historic continuity is essential where India is struggling for cultural identity and survival in the face of standardized values, mass produced culture, and alien influences. Conservation is a benefit to future generations, who will look back on their culture and history, as part of their background. Preserving our heritage thus would lead us to a secure, inclusive, and sustainable future; it is necessary for the social, cultural, and psychological equilibrium of mankind.

2 Conservation of Heritage in India

2.1 History of Conservation in India

The idea of conservation and transmission of values and knowledge systems are as old as our existence of human civilization. In today's date, we have numerous examples in physical and oral form, the results of the efforts taken by our ancestors to strengthen, maintain, and pass on the knowledge system to their next generation. The earliest form of written evidence available on efforts of conservation in India is from third century BC, where emperor Ashoka undertakes the conservation and protection of wildlife. There have been efforts by the Tughlaqs as well to conserve-built heritage. Although not available much as written records, conservation has been an integral practice in Indian civilization from long time.

The earliest effort for conservation of heritage in British India was the Bengal Regulation (XIX) of 1810 and the Madras Regulation (VII) passed in 1817. These regulations vested the government with the power to intervene whenever the public buildings were under threat of misuse. The Act XX passed in 1863 authorized the government to "prevent injury to and preserve buildings remarkable for their antiquity or for their historical or architectural value."

The Archaeological Survey of India (ASI) was established in 1861 in order to initiate legal provision to protect the historical structures all over India [1]. The Ancient Monuments Preservation Act (VII) passed in 1904 provided effective preservation and authority over the monuments and in 1905; for the first time, 20 historic structures in Delhi were ordered to be protected. At the time of independence, 151 buildings and complexes in Delhi were protected by the central ASI. At present, the ASI administers more than 3650 ancient monuments, archaeological sites and remains of national importance. These can include everything from temples, mosques, churches, tombs, and cemeteries to palaces, forts, stepwells, and rock-cut caves. The survey also maintains ancient mounds and other similar sites which represent the remains of ancient habitation (Archaeological Survey of India). In 1978, the State Department of Archaeology was set up in Delhi, but it lacks the power to acquire or protect buildings, and merely looks after those monuments de-notified by the ASI.

Apart from the ASI, archaeological work in India and conservation of monuments is also carried out in some states by state government archaeological departments. Most of these bodies were set up by the various princely states before independence. Indian National Trust for Art and Cultural Heritage (INTACH) is a registered society founded in 1984 to stimulate awareness for conservation of cultural heritage among the people. Apart from these, various states have and proposed laws for the preservation of their heritage, under which there is provision to notify heritage resources and include them as a part of area development plans.

2.2 *The Approach of Conservation in India and Its Impacts*

Among all the available legal provisions for protection of heritage in India, ASI has been the veteran organization in the field; majority of the protected cultural heritage resources in the country falls under its jurisdiction and its contribution in protection and conservation of cultural heritage resources is immeasurable.

The process of heritage conservation of the ASI starts with the "Identification and Listing" of potential cultural heritage resources. The boundary of the identified heritage resource is then demarcated, where the cultural property is studied carefully, and its extent is identified, viz. the immediate extent and the buffer zone. The immediate boundary is usually marked on the site by erecting boundary walls/ fencing around the identified extent. The concept of buffer zone [2] is to protect the potential extent of the site and to prevent undesired growth. After listing and studying the extent of site, the role of ASI is to maintain or enhance the quality of the site and surrounding, without disturbing the integrity and authenticity of the site, and to spread awareness among people regarding the importance and values associated with the heritage resource.

As per the national policy document for conservation, a conservation plan is mandatory before intervening into any heritage resource. This requires an in-depth study of the cultural property and its context, based on which conservation policies could be drawn. But while analyzing the approach of ASI toward conservation, it could be understood that it has been, by and large remained similar across the decades, although its conservation policy document has been updated with time. This gap between its theoretical framework and implementation impacts the heritage and its context in various ways.

2.2.1 Site Delineation—Meaning, Authenticity, and Impact

Fencing the heritage resource should be only an immediate/temporary response to prevent further encroachment or damage until the conservation plan is drafted. The actual problem arises when this demarcation of boundary becomes a mandate in every site. While not all heritage resources are intended to be separated from the public domain (which is evident from the absence of boundary in their original conception), there are some which comes along with boundary walls. Those heritage without boundary wall, once separated from the public or ticketed, loses the connect it was meant to hold and becomes alienated from the public.

While it is sometimes necessary to have boundary walls for the protecting the heritage, the design solutions for the same should also be well thought of. Inappropriate design solutions for delineating the site like metal wire fencing/solid stone wall create a psychological barrier—alienating the heritage.

Similarly, the treatment of landscape in the protected zone is often done by providing lawns, which requires regular maintenance, consumes lot of water, and gives no shade. Lawns are again adapted from a Westerner's perspective, where they required open front yards for the sunlight to fall during the winters, whereas shade giving trees have been the features of ancient Indian gardens, where the foliage cuts down the harsh rays of the sun, providing comfort for the tropical climate.

While there are indigenous ways of providing design solution for delineating the boundary and protected zones, heritage protection organizations in India like ASI, still follow the established westerner's style of design solutions, which creates a divide in the mind of the visitor with a clear line between "us" and "them."

2.2.2 Understanding the Socioeconomic Disparity Created by the Process of Conservation: Khajuraho, a Case Example

The neighborhood of the heritage resource is often ignored while building up conservation plans in our country; the context usually tends to suffer a socioeconomic disparity due to these myopic conservation plans. Let us take the example of the Khajuraho group of monuments falling in the region of Bundelkhand to understand this argument in detail.

The Khajuraho group of monuments is an excellent example of the Nagara style of temple architecture. It is an epitome of the art of space creation, spatial organization, and iconography. Khajuraho is one of the last known examples of North Indian temple, which is built in the indigenous Hindu architectural style. It is the manifestation of the expertise of the Jejakabhukti (later known as Bundelkhand) region during the rule of the Chandelas, in terms of building construction technology and human resource management; UNESCO recognized Khajuraho as a world heritage site in 1986 for its "human creativity."

The Chandelas/Chandela Dynasty ruled much of the Bundelkhand region of Central India between the ninth and the thirteenth centuries AD. The Chandelas are well known for their contribution to art and architecture, most notably for the temples at their original capital Khajuraho. They also commissioned several temples, water bodies, palaces, and forts at other places, including their strongholds of

Ajaigarh, Kalinjar and their later capital Mahoba [3]. This is an indication of the economic, social, and cultural richness of the Bundelkhand region that existed during their ruling period. But over the last millennium, the human developmental index of the Bundelkhand region (falls partly in present day Madhya Pradesh and Uttar Pradesh) has declined to an extent where it is now placed among the country's industrially backward regions; this is despite the fact that it is one of the "most visited" tourist destination of India. Although the revenue generated by the Khajuraho group of monuments has grown considerably, it does not reflect on the life of the local population. This disparity is created due to those vision documents that are generated without considering the vision and aspirations of the local population for their region. Ideally the vision document should have concentrated in the enhancement of quality of life of the local population in all terms, which could be achieved through capacity building and thus revenue generation.

Similarly, the natural ecosystem of the region has also been disturbed considerably. Bundelkhand, once known for its water resource management with its number of tanks, lakes, and ponds and for its dense forests, is left out with almost nothing and the region suffers scarcity for water in today's context. Once again the myopic vision plan for the area ignores the heritage of water engineering that existed which kept the region green; instead the protected zone and buffer zones are landscaped with neatly laid lawns and pavement after taking away the indigenous flora. This indicates that critical thinking is essential while drafting out interventions for heritage areas, especially when it is associated with resources which is associated with life of the public (Figs. 1, 2, 3, 4, 5 and 6).

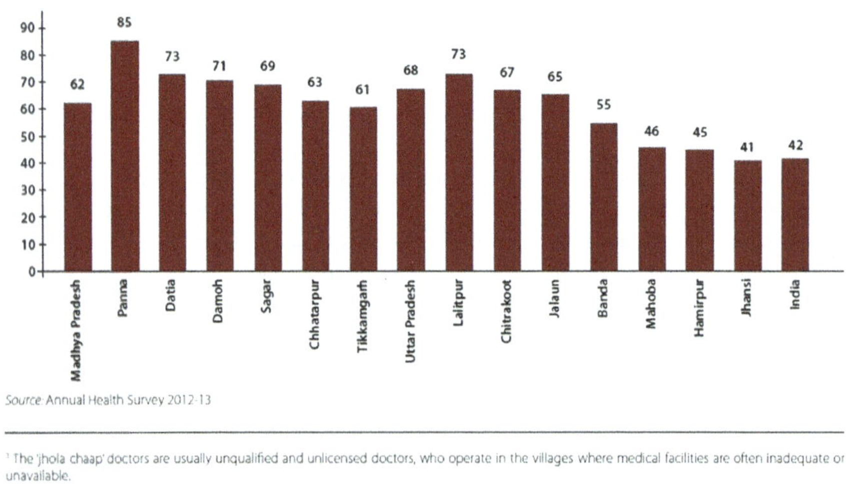

Fig. 1 Infant mortality rate 2012–2013

District	Literacy rate	Literacy index	NER (primary)	NER Index	Education index	IMR	Health index	Per capita income	PCI index	HDI	HDI rank
Jhansi	75.05	0.688	95.8	0.948	**0.775**	41	**0.621**	37999	**0.379**	**0.592**	1
Mahoba	65.27	0.566	100	1.000	**0.711**	46	**0.568**	34661	**0.336**	**0.538**	2
Hamirpur	68.77	0.610	94.2	0.928	**0.716**	45	**0.579**	26941	**0.240**	**0.511**	3
Jalaun	73.75	0.672	98.3	0.979	**0.774**	65	**0.368**	29476	**0.277**	**0.473**	4
Sagar	76.46	0.706	99.3	0.991	**0.801**	69	**0.326**	29028	**0.271**	**0.466**	5
Banda	66.67	0.583	100	1.000	**0.722**	55	**0.474**	24540	**0.200**	**0.465**	6
Datia	72.63	0.658	100	1.000	**0.772**	73	**0.284**	29307	**0.274**	**0.443**	7
Damoh	69.73	0.622	100	1.000	**0.748**	71	**0.305**	27999	**0.257**	**0.437**	8
Chattarpur	63.74	0.547	100	1.000	**0.698**	63	**0.389**	23799	**0.189**	**0.425**	9
Lalitpur	63.54	0.544	100	1.000	**0.696**	73	**0.284**	28989	**0.271**	**0.417**	10
Tikamgarh	61.43	0.518	100	1.000	**0.679**	61	**0.411**	21502	**0.148**	**0.412**	11
Chitrakoot	65.05	0.563	100	1.000	**0.709**	67	**0.347**	24011	**0.140**	**0.399**	12
Panna	64.79	0.560	100	1.000	**0.707**	85	**0.158**	23170	**0.178**	**0.347**	13
UP-Bundelkhand	69.30	0.616	98.3	0.979	**0.737**	56	**0.463**	29884	**0.280**	**0.493**	
MP-Bundelkhand	68.70	0.609	99.8	0.998	**0.738**	69	**0.326**	25893	**0.224**	**0.430**	
Bundelkhand	69.00	0.613	99.0	0.988	**0.738**	63	**0.389**	29021	**0.266**	**0.464**	
UP	67.70	0.595	94.2	0.928	**0.706**	68	**0.337**	26513	**0.149**	**0.397**	
MP	69.30	0.616	99.2	0.990	**0.741**	62	**0.400**	33028	**0.247**	**0.463**	
India	74.00	0.675	99.9	0.999	**0.783**	44	**0.589**	54042	**0.522**	**0.632**	

Fig. 2 Human development indicators, Bundelkhand and its districts

Districts	Literacy rate	Female literacy rate	Male literacy rate
Madhya Pradesh	**70.6%**	**60.0%**	**80.5%**
Sagar	76.5%	67.0%	84.8%
Datia	72.6%	59.4%	84.2%
Damoh	69.7%	59.2%	79.3%
Panna	64.8%	54.4%	74.1%
Chhatarpur	63.7%	53.6%	72.7%
Tikamgarh	61.4%	50.0%	71.8%
Uttar Pradesh	**67.7%**	**57.2%**	**77.3%**
Jhansi	75.1%	63.5%	85.4%
Jalaun	73.7%	62.5%	83.5%
Hamirpur	68.8%	55.9%	79.8%
Banda	66.7%	53.7%	77.8%
Mahoba	65.3%	53.1%	76.0%
Chitrakoot	65.0%	52.7%	75.8%
Lalitpur	63.5%	50.8%	75.0%

Source: Primary Census Abstract, MP and UP, 2011

Fig. 3 Literacy rates in the districts of Bundelkhand, 2011

As (%) of total

State/District	Single teacher schools	Schools with toilet for girls	Schools with Safe Drinking Water (SDW)	Schools with electricity	Schools with computers
Madhya Pradesh	**14.9**	**76.3**	**98.0**	**22.5**	**11.0**
Chhatarpur	11.6	86.5	99.5	23.6	11.6
Damoh	11.9	82.1	100.0	16.5	10.1
Datia	10.9	97.8	99.7	22.6	13.2
Panna	16.6	83.1	99.9	15.4	8.4
Sagar	14.0	97.0	100.0	20.0	9.9
Tikamgarh	9.0	95.2	100.0	13.9	6.4
MP-Bundelkhand	**12.3**	**89.1**	**99.8**	**18.5**	**9.7**
Uttar Pradesh	**9.0**	**81.3**	**98.5**	**34.8**	**7.5**
Banda	9.4	94.9	98.9	23.9	4.0
Chitrakoot	11.2	79.2	98.6	32.8	6.9
Hamirpur	3.8	94.5	97.7	35.1	11.9
Jalaun	20.7	73.7	95.1	38.3	3.8
Jhansi	4.8	99.7	98.4	65.2	12.6
Lalitpur	8.1	98.4	96.8	35.1	9.3
Mahoba	6.2	99.6	98.3	57.6	8.0
UP-Bundelkhand	**9.8**	**90.7**	**97.6**	**41.0**	**7.9**
Bundelkhand	**11.0**	**89.9**	**98.7**	**29.8**	**8.8**
India	**8.3**	**72.2**	**94.4**	**47.1**	**20.5**

Source: DISE 2011-12

Fig. 4 Infrastructure in schools 2011–2012

3 Heritage in the Indian Context

Any cultural property will have two attributes to it which shapes it up and gives it a meaning—a tangible part and an intangible part. The tangible attribute of the cultural property cannot be understood in isolation or in the absence of its intangible attribute. Both are interwoven so closely that the failure of one renders the other one lifeless.

On a global level, many cultural properties have undergone this separation and it is to save the last existing memory of the tangible, the property is conserved. Due to the cultural discontinuity, the sense of heritage in the western world largely ended up as conserving the remaining tangible aspect of the culture (even though the heritage remains incomplete). This practice has been carried on by the Europeans, and the same has been applied on to the Indian cultural properties too.

It needs to be understood that the sense of Heritage in the Indian context is way too different, different from the way the western perception of heritage. Indian

State/District	NER (Primary)	NER (Upper Primary)
Madhya Pradesh	**99.25**	**71.5**
Chhatarpur	100.0	73.3
Damoh	100.0	74.5
Datia	100.0	82.1
Panna	100.0	70.1
Sagar	99.3	74.9
Tikamgarh	100.0	77.1
Uttar Pradesh	**94.2**	**47.1**
Banda	100.0	64.1
Chitrakoot	100.0	81.0
Hamirpur	94.2	64.6
Jalaun	98.3	66.8
Jhansi	95.8	71.3
Lalitpur	100.0	77.2
Mahoba	100.0	72.3

Source: DISE 2011-12

Fig. 5 Enrollment in schools 2011–2012

Region	Percentage of women married before the legal age (18 years)		
	Total	Rural	Urban
Madhya Pradesh	**10.6**	**14.1**	**3.7**
Chhatarpur	25.4	31.2	8.6
Damoh	10.8	12	6.4
Datia	6.6	9.0	1.5
Panna	14.3	14.6	12.9
Sagar	8.2	11.2	3.2
Tikamgarh	27.1	30.6	14.3
MP-Bundelkhand	**16.2**	**19.7**	**6.8**
Uttar Pradesh	**5.4**	**6.7**	**1.6**
Banda	6.2	6.6	5.1
Chitrakoot	7.0	7.7	0.9
Hamirpur	2.6	3.1	0.8
Jalaun	4.7	4.5	5.2
Jhansi	3.8	5.6	2
Lalitpur	25.6	28.7	4.5
Mahoba	5.2	6.7	2.3
UP-Bundelkhand	**7.0**	**8.4**	**2.5**
Bundelkhand	**10.4**	**12.4**	**4.4**

Source: Annual Health Survey 2012-13

Fig. 6 Women who were married before the age of 18 2012–2013

civilization is one of the still surviving and continuing civilizations of the world. Unlike in the other global scenario, where events like invasions led to complete cultural destruction and disconnect, the Indian civilization continued to evolve despite of many such unfavorable conditions. The intangible attribute of heritage never got disconnected but evolved and continues to evolve. This holds true for many cultural heritage sites in India. The idea of History and Heritage in India is thus not to be something which is disconnected from the present-day society and frozen in its time. It is a story that is continuing due to its continuous cultural associations.

3.1 *Idea of Temples in Indian Context*

Temples in India, earlier, were spaces which were not just meant for worship alone; these were the institutions that bounded the community together. The temples served as learning centers facilitating knowledge and training in the fields like science and arts; a place for congregation and court of justice; places for discussions, debates, and art performances. Thus, the temples were the centers to which the community was anchored to, making themselves, an integral part of its community's life.

Secondly, the idea of deity/deities occupying the Indian temples is completely different from how it is perceived in other places of worship around the globe. In Indian culture, the temple is the "House of God" or the "Abode of God" (Devalayam)—where the deity resides. The deity here is not just representational or of an abstract form, but is a living entity. This manifestation starts with the process of Pranaprathishta (consecration) of the deity into the idol within the Garbhagriha. The deity then onwards is kept alive by carrying out the rituals and practices promptly as specified in the Agama Shastra. Any disruption caused to these routines will render the deity lifeless, gradually making the temple premises inactive.

3.2 *The Living and Non-living Temples of India*

A visit to the list of prominent protected temples across India would majorly reveal two contrasting scenarios:

- There are temples which are crowded with pilgrims, visitors/tourists, with the temple streets bustling with active street life throughout the day. These are the "Living temples" (e.g., Ranganatha Swamy temple—Srirangam, Madurai Meenakshi temple, temples of Kumbakonam, Vadakkunathan temple—Thrissur, Padmanabhaswamy temple—Thiruvananthapuram, etc.), where the deity has been kept alive through the strict regime of following the customs and daily rituals right from its consecration; due to which the community around has also been able to retain their connection associated with these temples, thus allowing the roots of culture to grow more deeper.

- And then, there are temples, even though exceptionally beautiful, monumental, and impressive, which stand in isolation with its context, totally inactive on its own. These are the "Non-living temples" (e.g., Khajuraho group of temples—Madhya Pradesh, Hoysala dynasty temples at Halebidu, Somnathpur), where the practice of daily customs and rituals has been interrupted due to various reasons (mostly because this part of India was continuously under attack by Islamic invaders), thus rendering the deity lifeless. The association between the community around and the temple becomes null, the moment the temple is rendered lifeless, thus leading to a gradual and complete disconnect between the temple and its people. Unfortunately, the lack of vision or lack of understanding of the basic principles of Hindu aesthetics has led to convert these temples into a dead institution, like a monument.

On the one hand, we have the cosmetically developed and maintained temples, where government agencies were able to preserve the architectural grandness, the outer skin, but not the soul of these social institutions. On the other hand, we have the living temples with interventions done over time, where community itself has managed to preserve the primary goal, i.e., keeping the Devis and Devtas in Devalayas alive.

This striking contrast noticed across examples of heritage belonging to the same typology, calls for the attention of all the stakeholders involved in the process of conservation of these cultural resources, to look back into the conservation process. The temples are potential centers of social life. Unfortunately, the lack of vision or lack of understanding of the basic principles of Hindu aesthetics has led ASI to convert these temples into a dead institution/monument. This raises certain critical questions toward the process of conservation of temples:

- What is desirable, a living temple with interventions made according to time or a non-living temple, which stands still with the golden stain of time, portraying all cultural and structural traumas it has gone through?
- Can the conservation project also aid in bringing the deity back to life, thus making the temple truly alive, by the process of re-consecration?
- Can there be a middle path where government institutions and the community could negotiate and resonate their ideas and aspirations?

4 The Way Forward—Integrated Approach for Heritage Conservation and Management

An ideal solution for heritage management should always be developed holistically by understanding the heritage through the three key parameters of people, place, and time, which means considering the natural, human, cultural resources (intangible part) associated with the heritage. The process of heritage management should be interdisciplinary involving professionals from various disciplines like conservation architects, environmentalists, artists, historians, legal bodies, other experts, and the local community, who are the various stakeholders of the heritage. An integrated heritage management plan is what is ideally needed, where the interests and role of all the stakeholders, viz. sectors/bodies (local, state, and federal) associated with the heritage area, the local interests and the legislations are taken into consideration for developing a sustainable management plan for the area. This involves looking at the development of a heritage area in a holistic manner which includes development through community/stakeholder participation and sustainable tourism development.

But the current system of conservation runs opposite to the historical/cultural narrative; once the heritage gets the protection labeled, it belongs to the state and not to the people. The insensitive interventions proposed under the label of heritage protection, separates the inseparable link between the built heritage and its community. This gap can be only bridged if there are strong partnerships built in the management plan, between the tangible and its intangible counterparts.

The primary stakeholders of any public heritage property are undoubtedly the people residing around the area. Their lives are usually interlinked in such a way that the intervention made to the heritage property affects their life as well. This right of belongingness should not be ignored while making conservation proposals for any heritage property. Rights-based approaches in heritage management instead of imposing a strict legal framework is what is required for public heritage resource, especially in case of resources like temples, where the rights of the community is vital for maintaining its life.

Effective conservation of a place is dependent on the level of support and understanding of all the inhabitants of the area. They are engaged as individuals in their communities, as neighbors in conservation, as well as farmers, foresters, businesspeople, and people working in local government and other public agencies. The conservation task is large. Effective partnerships between the department, people, and organizations can enhance the achievement of conservation outcomes by all parties. The participatory approach is thus an important aspect of heritage management, where heritage conservation is treated as a democratic process. In this process, the aspirations of the local communities are treated as an integral part in developing the management plan.

A range of opportunities to engage in conservation is provided by the legislation and policies. People can be nominated and appointed to local bodies for conservation and developmental planning. People are invited to comment on draft policies, management strategies and plans and proposed actions. People volunteer their time, skills, and resources to support conservation. All this engagement needs to be underpinned with general conservation awareness and educational activities.

5 Conclusions

Right from the existing education system, judiciary system and governing system, modern India has been built based on the borrowed concepts and ideas of the European society. This is the very reason why our society is still struggling and unable to find the missing links to bridge the gap between its past, present, and future. The institutions responsible for heritage protection in India look at the Indian Heritage resources through the European perspective. They isolate the property from its people and uproot it from its context. Indigenous problems require indigenous solutions; not those solutions that are borrowed from an alien context.

Also, the sense of heritage and history in India has a different meaning altogether, when compared to the Western ideology. Hence, the approach toward its conservation also needs to be tailor-made according to its demands. No changes to the existing legal tools are forth coming, either through disciplines, scope, or processes. India needs its own decolonized theory to inform heritage practice which protects material and values and envisions a future co-envisioned and co-created with the people to who the heritage belongs.

Understanding the monument, site and its context before intervention is a very crucial step for the process of conservation of any cultural heritage. Each heritage comes with its own nature, complexity, as well as extremely complex relationship with its people. Understanding this interrelationship gives an insight into the actual demands of the heritage as well as its people. The current process of heritage conservation in India disconnects the people from the heritage under the pretext of "protection of monuments," leading to spatial and temporal isolation of the heritage.

Heritage management without a community leads to disconnect from historical and cultural narrative and thus to social exclusion. To bring about sustainable development in the process of heritage management, the heritage should be integrated to the life of the surrounding neighborhood. The aim of the conservation process should be for enhancing the existing goodness/quality of the heritage and its context, not to reduce the bond by restricting their interaction. Conservation and management policies should focus on mutual advantage where the community and the heritage support each other in quality improvement. The consultation process and the democratic, participatory processes are critical to draft such plans and policies and that is what we need to fight for.

References

1. https://en.wikipedia.org/wiki/Archaeological_Survey_of_India
2. Li Q, Yuichi F, Morris M (2014) Study on the buffer zone of a cultural heritage site in an urban area: the case of Shenyang Imperial Palace in China. The sustainable city IX, vol 2
3. Dahiya PD (2017) Ancient and medieval India: for UPSC and state civil services examinations. McGraw Hill Education (India) Pvt Ltd, Chennai

Social Inclusivity

Scientific Ideation Towards Visionary Development Strategies for Indian Urban Environments

N. Devi Prasad

Abstract The aim of development projects encompassing the urban built environment is to ensure a better quality of life for citizens in the present and for the foreseeable future. Broadly, these involve redevelopment, conservation and renewal in the brownfield realm and infrastructural improvements and new area developments in the greenfield sector. The process involves understanding the needs of the built environment and proposing interventions through a process of dialogue among city planners, urban designers, architects, financial planners, government officials and public and private stakeholders. While most public agencies rely on precedent while envisioning new development projects, visionary thinking often evolves from sustained intellectual effort. The aim of this paper which is aligned to the Indian context is to establish a case for: (i) Inclusion of academics and researchers in the consultative process towards a more intellectual framework while arriving at solutions to issues confronting the built environment, (ii) Creating a dialogue among institutional researchers, government development bodies and people's representatives towards mutual understanding of concerns, (iii) Facilitating the access of researchers to administrative processes and operational logistics to enable teaching universities to provide more relevant education to students, (iv) Sharing research studies with urban development agencies to provide ideas for possible implementation. The importance of academic inputs in large developments projects is highly significant. Any large projects such as new town developments, area redevelopment projects and urban renewal initiatives need a significant amount of intellectual application before design proposals are envisaged and executed.

Keywords Built environment · Scientific ideation · Development authorities · Urban development · Academics

---

N. Devi Prasad (✉)
School of Architecture, Vellore Institute of Technology, Vellore, India
e-mail: ndeviprasad.ams@gmail.com

K. Thirumaran et al. (eds.), *Sustainable Urban Architecture*,
Lecture Notes in Civil Engineering 114,
https://doi.org/10.1007/978-981-15-9585-1_7

1 The Institutional Process of Urban Development in India

A study of the major urban development authorities in India—Chennai [1] Mumbai [2] Kolkata [3], New Delhi and Bengaluru—reveals that all of them have over ten members on their executive board. The committees comprise members of the bureaucracy related to urban development and finance, the chief architect and town planner, elected people's representatives, technical heads of infrastructural agencies related to civil works, water supply, sanitation and electrification and a few invited members. Fundamentally, these institutions are comprised of executive members of the development agencies of the government entrusted with the creation and realisation of development plans through a process of perceived needs and public opinion. It may be reasonably inferred that based upon the calculated wisdom of these members; many environmental interventions are introduced.

The members in charge of the operational logistics work on the premises of these committees to produce tangible suggestions for urban improvement to arrive at final proposals. While this process is seemingly complete, the ideation skills of these committees are impeded by their preponderance with daily responsibilities and projects under execution and real or imagined bureaucratic restrictions. It is common to see youngsters who join government departments filled with an energy born of idealistic and youthful enthusiasm disappointed by their rigid working environments. Thereafter, they choose to either take their intellectual wares elsewhere or become more bricks in the wall and sink into the thought patterns of their surroundings.

Political directives prevail on the final directions of several development projects, and while some guidance is sought from public discourse, they are often based on limited assessment of options. Even when large-scale projects are publicised for selection of consultants and partner developers, the basis and terms for these requests for proposals (RFPs) are often not clearly established by data or precedent wisdom. This does not necessarily imply lack of application of mind but simply lack of awareness of multiple strategic viewpoints when assessing possibilities for development. In the case of smaller Indian towns, especially those which have populations of 100,000 people upwards and vested with municipal corporations, development activity is vested in the hands of a few technical personnel, bureaucrats and elected representatives. Critical decisions are made which affect future urban growth and cause problems already experienced by other cities in transition. The intellectual content for any proposals for city expansion, land use layouts, transport alignments, new buildings and architectural regulations is usually provided by local resources and there are serious capacity limitations on this front.

While some projects may prima facie appear logical to execute, as for example, the creation of a commercial hub in a city centre area to capitalise real estate, the options available to achieve a mix of social and commercial objectives coupled with precedent research may achieve substantially higher socioeconomic benefits for the costs involved. Developers use professional consultants to prepare project plans and details and while these entities are well suited to carry out the work with substantial agility; they are invariably incapable of research wisdom and analysis which establish the most efficacious methods of operation. For instance, commercial activities in an Indian traditional market are driven by market driven factors and a symbiotic relationship exists between the buyer and the seller. Aggressive development of such commercial activity without understanding such dynamics often leads to failure of the spaces to meet the functions for which they were intended such as the Lily Pond Complex in Chennai [4] and the new KR Market building constructed in 1990 at Bengaluru. Another proposal which would have benefitted from scientific ideation is the Mass Rapid Transit System connecting north and south west Chennai which suffers from serious connectivity issues and lack of vision while creating commercial spaces [5]. Any proposal to intervene in such an environment must include a larger understanding of the market within the overall framework of city functions and user [6].

Routinely, the Housing Board Authorities in Indian states undertake large housing projects intended to serve various categories of affordability levels. The design briefs for many of these projects need to be closed before design drawings of planning layouts and housing are finalised. Urban design plays an important role in the qualitative aspects of public space and the architectural content needs to address the needs of sustainability and growth. The foundations for these concepts could be enhanced through meetings with academic professionals from institutions of higher learning. Similarly, the Public Works Departments of states deal with hundreds of civic interventions on a regular basis—these include corporation offices, public toilets, public feeding centres, utility buildings, parks and playgrounds and many other entities which play an important role in defining the streetscape of any urban settlement. Design and planning content of such developments leaves much to be desired.

2 The Thinker and the Doer and Everybody in Between

Academic institutions can contribute in a large way as resources to provide answers to questions that are not apparent. The process of academic learning presumes an unbiased approach to understanding development processes through the multiple prisms of precedent examples conveyed by research techniques and future implications through mathematical modelling, statistical analysis and multiparametric projection efforts. The access of researchers to knowledge bases is uniquely different from those of professional practitioners. While a consultant or professional would be able to understand the operational logistics of any built development project, an academician would be able to dispassionately analyse the impact of such development or question even the need for such development. Deep scientific thinking is the forte of academic ideation and substantive inputs can be gleaned through such rigour. It would be easily conceivable that the best-intentioned plans of policy makers and professionals could be overturned by serious academic thinking based upon researched facts. While the government and public authorities are invested in the provision of infrastructure to benefit public interest, available intellectual resources are often outdated and research cells are inadequate.

While the priority of urban development authorities lies in the realisation of plans, academic institutions by their very nature immerse themselves into comparative studies and intellectual mathematical analysis of simulated situations and arrive at purist or logical solutions devoid of political or impression-based inferences. While academics need to learn the governance methodology of planning bodies, they are not primarily shackled by external influences in the process of their research. If for example a local area development plan was to be proposed, even the quality of the decision-making process to identify the specific area that needed attention would be enhanced with inputs from the research community who could provide more scientific reasoning for their choices as a matter of pure rigour which is part of their routine process of enquiry, analysis and conclusions.

While professionals are limited by the constraints of time and finance, they are more equipped to deal swiftly with changes in ground realities. They are also more invested on a daily basis with the daily progress of projects from conceptualisation to realisation entailing an emphasis on product rather than process. Policy makers too are burdened with realisation of projects in order to meet infrastructural needs. Process is sometimes a token gesture to satisfy administrative protocols than devotion to achieving the right solution. The inclusion of academic ideation in the process is a double-edged sword in that while on the one hand it unleashes the potential of process-based thinking, it is often seen as being more of a speed break to fundamentally sound projects. There is also the agility of academics to deal with rigorous timelines and schedules. While this may be attributed partly to the institutional responsibilities of teachers and researchers, it can also be traced to the greater emphasis of learning institutions on pure rather than applied research. Academicians are thus more rigorous at process-based thinking but less knowledgeable of critical time and finance constraints while project controllers are more focused on time-bound realisation of projects within specified financial constraints. The appropriate approach to any development project involving large financial outlays and extensive public impact would be to involve academic institutions and especially students to create the background logic for the project and provide directional inputs based upon thorough studies based upon which professionals may take forward detailed initiatives to completion.

3 Infusing Institutional Academic Ideation into Urban Development Plans

One of the laments of academic institutions is their lack of connectivity with industry and real-time issues. Architectural academics pursue an agenda of knowledge delivery based upon syllabuses created at different points in time. These syllabuses are revised usually once in five years or sometimes more frequently if there is some serious inconsistency recognised in the teaching process. In the current context of globalisation and rapid technological advancement, this period is too prolonged, and as most of the subjects are related to theories, there is a serious lack of connect with project development practices. Token subjects are offered to students to offer proficiency in practical applications but these are few and far between and are often dated in content.

One important aspect which was pointed out by a practising professional is that the process of project definition, funding methods and related bureaucratic linkages are not well known to the general public. There are special purpose vehicles (SPVs) for projects, infrastructure development funds, civic maintenance funds and various other heads under which projects are executed. The process of sanction of projects, clearances and approvals is well known only to those who are in the midst of project sanctions and approvals on a regular basis. In this scenario, the academic is lost in a sea of bureaucratic acronyms and prefers to retreat into the relative comfort of theoretical certainty.

Accreditation norms of most universities credit professional engagement or consultancies as desirable demonstrations of a tangible connect between the realms of education and professional practice [7]. If institutions were to be given opportunities to engage in public built environment projects, they would experience first-hand the realities of governance and execution and both students and faculty would be able to appreciate the need for expedience in project execution. Government policy makers would be able to access intellectual thought at reasonable cost and adopt ideologies as they deem relevant to the context. Students and faculty members who participate in data analysis, physical and socioeconomic surveys too would realise their learning outcomes in the field and be rewarded in the form of subject credits and certified acknowledgements of their achievements.

Public participation in government projects is an integral part of democratic governance. The implementation of such participation is effected through the conduct of social media messaging, posters in public offices and field surveys of affected populations besides public meetings and organised interactions of representative leaders of groups. These are essential to garner public support for development projects to succeed. However, while assessing the feedback and suggestions of the general public and sections of intelligentsia, the analysed outcomes and benefits/disbenefits of projects may not be apparent to the stakeholders. This gap can be addressed by research-based intelligence (Fig. 1) [8] that can be offered by teaching institutions. Even while requests for proposals are floated and professional consultants are appointed, these consultants are not primarily equipped to conduct feedback analysis or detailed technical justifications for many of their solutions. Conceptual solutions backed by data are best obtainable from educational faculty resources and students and many professional offices invite niche surveyors and educational institutions to assist in data collection. The management, analysis and conclusions are however more difficult to expect from statistical surveyors. Professional data gatherers can elicit volumes of information without specific outcomes being realised and can also be influenced by premeditations.

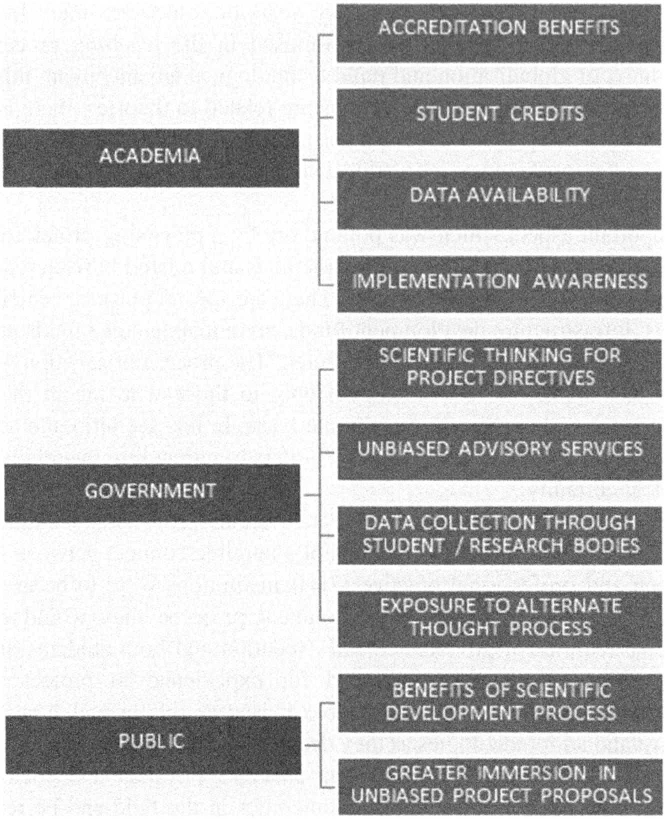

Fig. 1 Scientific ideation towards enhanced project outputs

4 The Policy Maker–Academic Researcher Handshake

The more continuously that academia interacts with real-time proposals, the more enriched and intellectual yet pragmatic would be the contributions of successive engagements. Over a period of time, researchers would be equipped with enhanced agility and operational skills but with the freedom of liberal thought. The gap that exists between practitioners and academics could thus be bridged through multiple iterations of teachers and students with policy makers.

The academic thesis is a ubiquitous feature of all academic activity in undergraduate and graduate education in India. While the undergraduate thesis synthesises the information that has been learnt over the first few years of the programme, the graduate thesis seeks to be more far reaching in its scope by opening up new vistas for research and an initial foray into the research methodology process. The graduate thesis process is of particular interest here as it entails a considerable amount of deeper thinking including a demonstration project often in a realistic setting.

In the case of architectural theses, a case is sometime built up for a particular form of development proposal in a neighbourhood or a city level intervention or a solution to a government sponsored live proposal. At the graduate level, a thesis in urban design may take the form of an area development proposal or policies at urban district level. A planning-oriented thesis could research into the dynamics of city growth, land use and development regulation parameters. However, one views the final product of any of these subjects; the process is substantially rigorous. For example, in the case of an architectural intervention, a great deal of research is invested in building programming, analysis of regulations, relationships with surrounding physical elements, site analysis and climate and energy studies. In the case of a thesis on the subject of urban design, field studies are conducted rigorously encompassing land use, built volumes, contextual impacts, heritage assets, socioeconomic indicators and infrastructural realities besides public surveys and popular perception analyses. Planning studies involve a huge volume of data gathering and compilation from several private and government inputs and related analysis and interpretation assiduously carried out over a period of several months.

The common thread tying these academic pursuits is the pursuit of their possible application into reality, even if substantial parts of their emphases are on theoretical or conjectural ideas. The research in the field and the investigation of the subject through virtual means and direct engagement with private professionals or public servants are carried out in an environment largely devoid of external influences of a bureaucratic nature. In a sense, theoretical studies carry a seed of creativity and lateral thinking that executing bodies often do not have the luxury to exercise as a result of their constant engagement with the "to do" approach over the "to think multioptionally" approach. This does not imply that policy planners are themselves not endowed with vast knowledge born of experience and familiarity with realistic situations, but are not stakeholders in research-based approaches to development model thought.

There are over four hundred institutions in India imparting planning, urban design and architectural education and a vast body of applicable field knowledge is generated constantly. The reality is that a vast amount of research conducted in academic institutions stays within research circles and student project archives and do not trickle into policy planning and implementation at the ground level. A case for involvement or sharing of knowledge by reputed academic institutions with the government on various projects of researchers vested in the archival reservoirs of government will prove to be valuable resources for policy planners and development authorities.

In a report on The Foundation Role of Universities as Anchor Institutions in Urban Development by Debra Friedman, David Perry and Carrie Menendez (as part of USU—the Coalition of Urban Serving Universities, USA), the authors present a

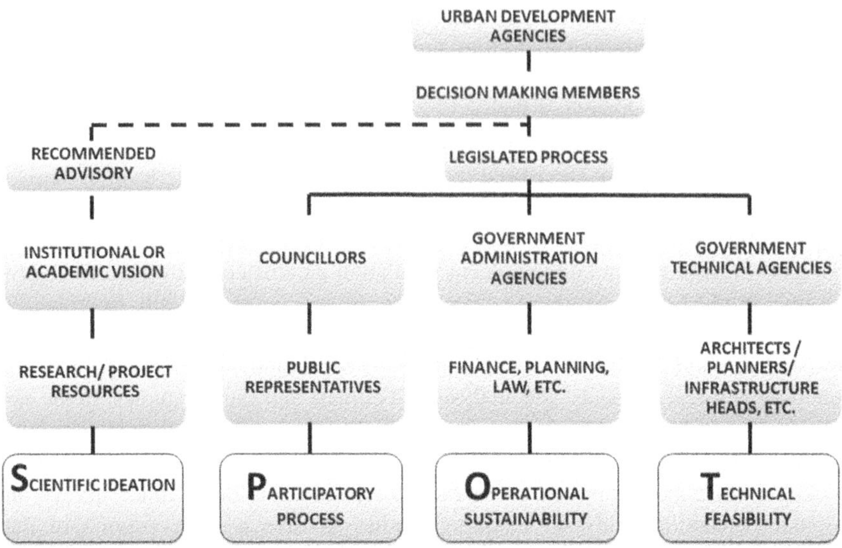

Fig. 2 The SPOT urban development policy process

compelling case on the role of academic institutions in urban development and say that "it will require more than a single snapshot survey to make the case that urban serving universities are important to the wellbeing of their cities" [9].

The government needs to encourage efforts of academia by providing focused access to information and government databases pertaining to the interests of researchers and positioning of academic intellectuals on the panels of infrastructural projects besides encouraging officials to participate in the relevant programmes of institutions which deal with built environment research. There is scepticism regarding the valuation of the research that is carried out in several institutions of higher learning but as there is already a framework in place identifying the categories of institutions according to their rigour in academics and governance, efforts to rope in those institutions which have undisputed credibility would strengthen the thought processes which lead to urban development projects. Figure 2 [10] shows the SPOT Urban development policy process.

The irony is that many studies of great merit, that could have fuelled actual field application, lie wasted by the wayside after their academic needs have been fulfilled. A public initiative is already available in the field of biotechnology [11] for conversion of research into products. Similarly, a review of student and faculty ideation efforts through a systematic dialogue-based interaction among stakeholders at regular intervals would yield positive results in the realm of the built environment.

5 Action Plans for Implementation

Some positive ways to create greater consensus and visionary direction to urban development projects would be to

(a) Facilitate regular events of learning and sharing of intellectual pursuits with policy planners at local and regional levels.
(b) Include teachers of urban design and planning in awareness programmes on governance, bureaucracy and commercial process while dealing with projects and brief institutions on pressing needs of society to enable further research in directions effectively reaching the community.
(c) Revise educational curricular content to meet projected local, regional and national needs.
(d) Provide space for institutional academic advisors as a matter of regular procedure in urban development planning at all levels.
(e) Involve students as project associates on a regular basis.
(f) Liberate non-sensitive government data relating to planning, socioeconomics, infrastructure and land mapping for research use.
(g) Create space for an advisory chair on public development projects mooted by municipalities and urban development authorities. This could be done by creating a data base of universities offering expertise in planning and architecture in relation to their accredited competence as measured by their academic scores in assessing agencies like National Assessment and Accreditation Council (NAAC) and allowing participation from those sources on a rotational basis. These institutional advisories could help in framing guidelines for planning of new towns and layouts or large redevelopment and infrastructure projects at a metropolitan level and architectural and local planning strategies at the level of small and medium sized municipalities

A cross exchange of this nature would facilitate greater empathy of thinkers with the challenges of execution of development models and provide government officials an opportunity to understand intellectual thinking and process importance while formulating urban interventions. The democratic process would be strengthened by such involvement and the greater involvement of intellectuals into the development planning process would benefit public interest at large. The cynicism that accompanies secrecy of project formulation would also be reduced if consultative and collaborative processes were employed more proactively.

References

1. CMDA (n.d) The information hand book under right to information act. Available at http://www.cmdachennai.gov.in/pdfs/rti_1_to_7.pdf. Accessed 21 Dec 2019
2. MMDA (2013) Administration. Available at https://mmrda.maharashtra.gov.in/administration . Accessed on 21 Dec 2019

3. KMDA (n.d) Kolkata metropolitan development authority—Orgonogram. Available at: http://kmdaonline.org/home/org_chart. Accessed 21 Dec 2019
4. Osborne J (2013) Why Chennai's vendors prefer the sidewalk to the mall. In: The rockfeller foundations: informal city dialogues. Available at https://nextcity.org/informalcity/entry/chennais-lilly-pond-street-vendors-fight-for-an-outdoor-market. Accessed on 10 Jan 2019
5. Madhavan N (2010) Mass rejected Transit system. In: Business today magazine. Available at https://www.businesstoday.in/magazine/case-study/mass-rejected-transit-system/story/5396.html. Accessed on 10 Jan 2019
6. Global-is-Asian staff (2017) Cities for citizens, by citizens: public participation in urban planning. Available at https://lkyspp.nus.edu.sg/gia/article/cities-for-citizens-by-citizens-public-participation-in-urban-planning. Accessed on 18 Jan 2019
7. NAAC (2019) Institutional accreditation manual for self-study report universities. Available at http://www.naac.gov.in/images/docs/Manuals/University-Manual-29th-August-2019.pdf. Accessed on 18 Jan 2019
8. Prasad Devi N (2020) Scientific ideation towards enhanced project outputs. Created by author on 19th Jan 2020
9. Friedman D, Perry D, Menendez C (2013) The foundational role of universities as anchor institutions in urban development a report of national data and survey findings. Available at https://usucoalition.org/images/APLU_USU_Foundational_FNLlo.pdf. Accessed on 18 Jan 2019
10. Prasad Devi N (2020) The SPOT urban development policy process. Created by author on 19 Jan 2020
11. BIRAC (2020) Promoting academic research conversion to enterprise (PACE). Available at https://birac.nic.in/desc_new.php?id=286. Accessed on 18 Jan 2019

Social Inclusivity: A Case Study on Community Resilience on Kerala Flood-2018

Sameer Ali and Abraham George

Abstract Urban resilience for any city is a gap to be fully understood and assimilated in urban planning. Globalisation and rapid urbanisation in recent years have led to newer challenges as higher densities, greater demand for infrastructure, resources, environmental and man-made hazards are on the increase. Cities are trying to cope up with the rising needs through various planning techniques and modern applications and planning models. However, each city's landscape is different and many a time, these approaches might not adhere appropriately to every aspect of a city. One of the worst situations where the resilience of a city would be truly realized, is when disaster strikes. The flood in Kerala State in August 2018 is one such example where researchers can study a lot. Not only the fact of 'coping up to the maximum damage possible', but it is also 'how fast human lives can be brought back to their normal stable level'. Once pre-disaster state is achieved, the need to improve services or to continue as usual is another question to be resolved. In most cases, the governmental and other agencies would strive to attain the lowest acceptable condition. However, it is most appreciated if the resilience exceeds the original level with new approaches to planning, design and infrastructural capabilities so that the disaster, even if it strikes again will affect the damage prevention better. The ability of the city or the extent of resilience shown by Kerala is strong, especially when compared to similar scale disasters that have struck in India and even other parts of the world. The paper tries to study and evaluate these factors that lead to faster resilience of Kerala's state to model a flexible and more effective urban resilient planning approach.

Keywords Flood · Disaster · Community resilience

S. Ali (✉) · A. George
Indian Institute of Technology Kharagpur, Kharagpur, West Bengal, India
e-mail: sameer_aaa@live.com

A. George
e-mail: profageorge@gmail.com

K. Thirumaran et al. (eds.), *Sustainable Urban Architecture*,
Lecture Notes in Civil Engineering 114,
https://doi.org/10.1007/978-981-15-9585-1_8

1 Introduction

A simplified version of resilience would be simply *'being strong'* or *'bouncing back'* after a fall, or workers' capacities to manage stress, as per the former chancellor, the bank of England, George Osborne's, discourses around building a *'resilient economy'* [52]. Resilience also means *'to hold on to one another'*, for greater strength in unity which is seen in the Kerala floods of 2018. Recommendations for increasing urban resilience include techniques 'to multiply weak ties and to form alliances with people of different faiths' and to mix housing patterns and provide adequate services in poor neighbourhoods [55]. In short, *a multidisciplinary form of urban planning* or *mixed residential system* is seen appropriate in bringing the resilience level up.

Finding the appropriate predictors of resilience has become an arduous task because of the high-rising willingness to see the behaviours of people as influenced by potent exchanges with the environment and significant others. *Cross-legged* prospective studies are necessary with advanced statistical methodologies and multivariate data collection strategies capable of analysing transactional data [23]. Urban resilience has been of paramount importance for cities around the globe especially, mega-cities *which exceed 10 million population* [26]. Of the 23 mega-cities in the world, 16 of them are in coastal zones and within 100–50 km elevation from the MSL [26]. Research also indicates that there is a critical gap between urban and coastal research, which is to be done.

It is often observed that urban resilience lacks proper cognisance by authorities, in the political and economic aspects, thereby bringing 'social injustice' and 'undesirable environmental loading' [9]. As such, at times of crisis a city is usually not well equipped to handle extreme situations. However, it is only during post-disaster events; significance of a resilient city is taken into consideration.

1.1 Urban Resilience- Indian Context

India is a developing country with the highest number of cultural diversities in the world. This would mean eventual transition to a mixed cultural process governing colonial urban development that is unique and discrete specific to each region. This is seen in colonial cities where utmost importance is given to values, behaviour, traditions and the distribution of social and political power within it [21]. Studies have proven people's emotional connections to their homeland contribute positively towards environmental concerns [10]. Such is the case of Kerala during its devastating flood in August 2018. The local response was spontaneous and incertitude led to the best form of empirical support that is crucial people's response in any disaster.

Failure of a society to cope up with a disaster or a disruption simply corresponds to *their 'inability to augment in a sustainable manner to manifest resilience'* [31]. Social attitude plays an explicit role for any society, and they are the first responders when a disaster strike. The influence they can create in increasing resilience, and as such reduce the effect of a disaster and its mitigation, is definitely indispensable. The faster and appropriate the response of 'people as sensors', higher is the positive impact on resilience [24]. The tech-savvy segment of the public helps to create local and global awareness, first-hand reports, upload pictures/videos, create blogs helps promulgate information sometimes even faster than government agencies to help seek out family, friends or any needy person. However, the spread of unauthenticated fake information is also a concern as it spreads avoidable panic in an already adverse state.

2 Kerala

Kerala is a south Indian state with more than 33 million populations, with a geographic area extending to 39,000 km^2. It has a coast of 590 km, and the width of the state varies between 11 and 121 km. Geographically, Kerala can be divided into three climatically distinct regions: the eastern highlands; rugged and cool mountainous terrain, the central mid-lands; rolling hills, and the western lowlands; coastal plains (Fig. 1) [13].

Kerala ranks first among all the other states in terms of Human Development Index (HDI) [36]. Contrary to many other states of India, Kerala ranks very high in many other Human Development Indicators as well at par with those of developed countries. For instance, Kerala ranks first among states in inequality adjusted HDI which simply corresponds to least loss of HDI in terms of inequality [41]. As compared to the National average of 74% literacy rate, Kerala reports a rate of 94% 30]. The higher the HDI, the higher is also the coping capacity but with greater cumulative loss potential, it might pose an even higher risk.

About 90% of the rainfall in a year occurs during six monsoon months, from June to November. Kerala State has an average annual precipitation of about 3000 mm (Fig. 2) [15]. The continuous and heavy precipitation that occurs in the steep and undulating terrain finds its way into the main rivers through innumerable streams and water courses. Kerala being a coastal city with a coastline of 590 km is however vulnerable to Natural Disasters such as Floods, Tsunamis and even Landslides along the slopes of the Western Ghats. There are 39 hazards categorised as naturally triggered hazards and anthropogenic ally triggered hazards by the *Kerala State Disaster Management Plan* [20]. Also, Kerala is one of the most densely populated states of India with 860 persons per sq.km making it more susceptible to losses and damages from a disaster [30].

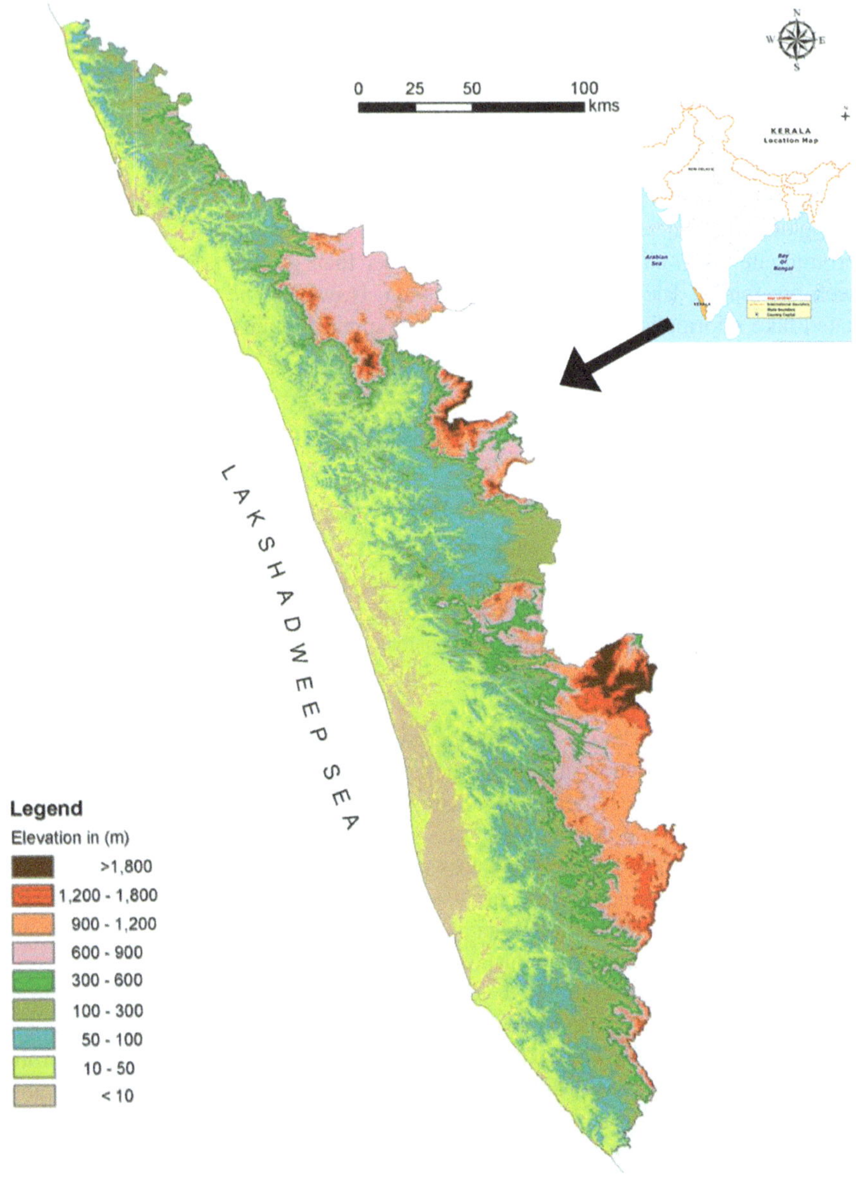

Fig. 1 DEM, digital elevation model of Kerala [61]

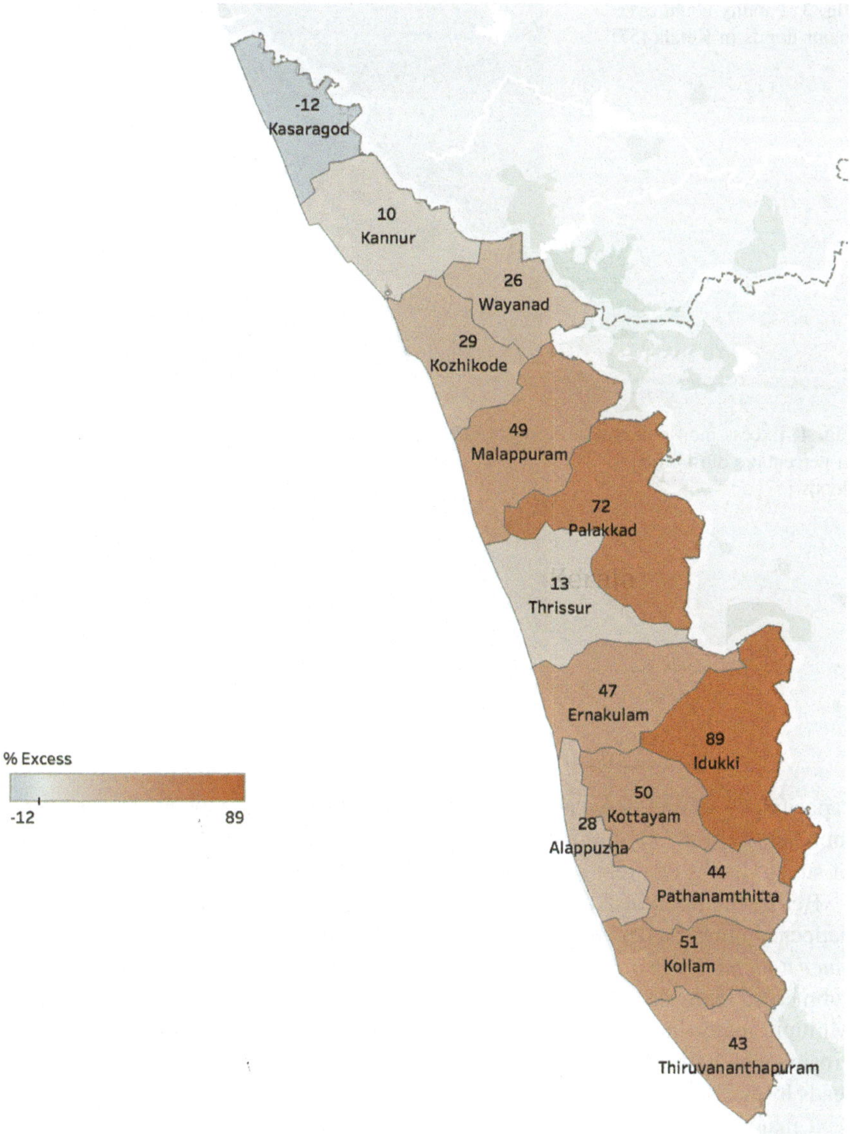

Fig. 2 Excess rain percentage Kerala for monsoon 2018 [15]

2.1 Kerala Floods

Out of the above-mentioned hazards, floods are the most common of natural hazards in the state. Close to 14.5% area of the state is prone to floods. Riverine flooding is a recurring event consequent to heavy or continuous rainfall exceeding the percolation

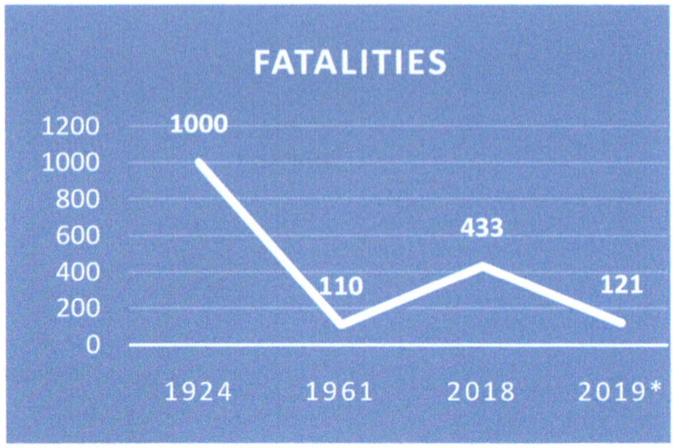

Fig. 3 Fatality count over major floods in Kerala [57]

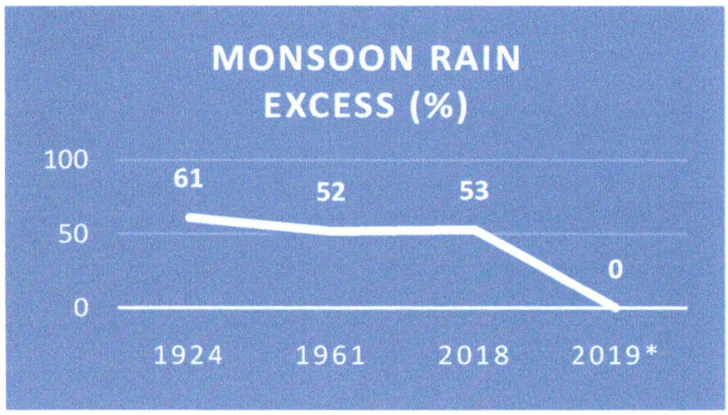

Fig. 4 Excess monsoon rain in percentage during major floods [57]

capacity of soil and flow capacity of streams and rivers. The events that trigger an inundation are mainly rainfall, channel slope, land use in the flood plains, materials of stream banks and relative height of banks [17] (Figs. 3, 4, 5 and 6).

Before the flood of 2018 and 2019, the previous flood of such devastating scale happened almost a century ago in 1924. More commonly known as the '99 floods; *since it happened in the Malayalam calendar year of 1099*, the then flood had deeply submerged many districts of Kerala from *Thrissur* to *Alappuzha* even parts of *Idukki*. Multiple major landslides were triggered in *Karinthiri Malai* probably due to toe erosion which irreparably damaged the then *Munnar road*. The present-day road that leads from *Ernakulam* to *Munnar* was constructed after this incident. After 1924, the next major flood was in 1961 in the Kerala *Periyar* Basin with a 52% increase in Monsoon rains during that time. Casualties were relatively lesser; *at 110*, this time as compared to the previous event due to lesser urban densities in these areas.

The flood seems to be recurring at a closer frequency especially when it is happening back-to-back years. The year 2019 ran through a deficit of 29% rainfall on 1 August 2019 to no deficit on 14 August 2019 which clearly implies the excess rain received in just 2 short weeks [17]. 40 people have been injured in flood-related incidents while over 21 people are believed to be missing in the state after heavy flooding. A total of 1789 houses have been completely damaged due to flooding, and over 26,000 people have taken refuge in relief camps [17].

Fig. 5 1924 Flood level marking at Puthiyakavu temple, Ernakulam [58]

Fig. 6 A view from Munnar after the 1924 floods [59]

2.2 August 2018 Kerala Flood

From June to August 2018, Kerala has received one of the heaviest rainfalls, 42% more than the normal average and thus experienced the worst of floods in its history since 1924 [53]. With the heaviest rainfall from 1–20 of August, the state has received 771 mm of rainfall [20]. The rains triggered several landslides in different districts and the forced the release of excess water from over 37 dams further reportedly aggravated the flood situation. From 10 districts, 341 landslides had been reported during this period. The most intense of rains happened on August 15 and 16 with intensity of 235.5 mm which has a probability of 0.5% chance for any given year. It only rose to 294.2 mm on the third day August 17 which is unparalleled [35]. The devastating floods and landslides affected 5.4 million people, displaced 1.4 million people and took 433 lives out of which 268 were men, 98 women and 67 children [57].

Six of the major reservoirs in the state were already at 90% capacity even before the rains started lashing in August 2019.[1] The untimely release of water from the dams was also said to be a reason that attributed towards the flooding [35]. The water levels in several reservoirs were almost near their full reservoir level (FRL) due to continuous rainfall from 1st of June 2019. Another severe spell of rainfall started from the 14th of August and continued till the 19th of August, resulting in disastrous flooding in 13 out of 14 districts. Over 175 thousand buildings have been damaged either fully or partially, potentially affecting 0.75 million people. More than 1700 schools in the state were converted into relief camps during the floods. The worst affected were the workforce in the informal sector that constitute at least 90% of Kerala's workforce [28]. It is estimated that nearly 7.45 million workers, 2.28 million migrants, 34,800 persons working in micro-, small and medium enterprises, and 35,000 plantation workers, *majority being women*, have been displaced. The major issue being, these workers are usually daily wagers, and they experienced wage losses for 45 days or more (Fig. 7; Table 1).

2.3 Analysis of 3 Days Cumulative Rainfall of 15–17 August 2018

To analyse the August 2018 flooding phenomenon of Kerala, daily rainfall data from 1 June 2018 to 20 August 2018 has been obtained from the Indian Meteorological Department (IMD). On scrutiny of data, it has been found that cumulative rainfall realised during 15–17, August 2018 was significant, with places such as *Peermade* receiving more than 800 mm of rainfall [15] (Fig. 8).

[1]Data as on 20th August 2019, the flood condition still prevails as on 31 August 2019.

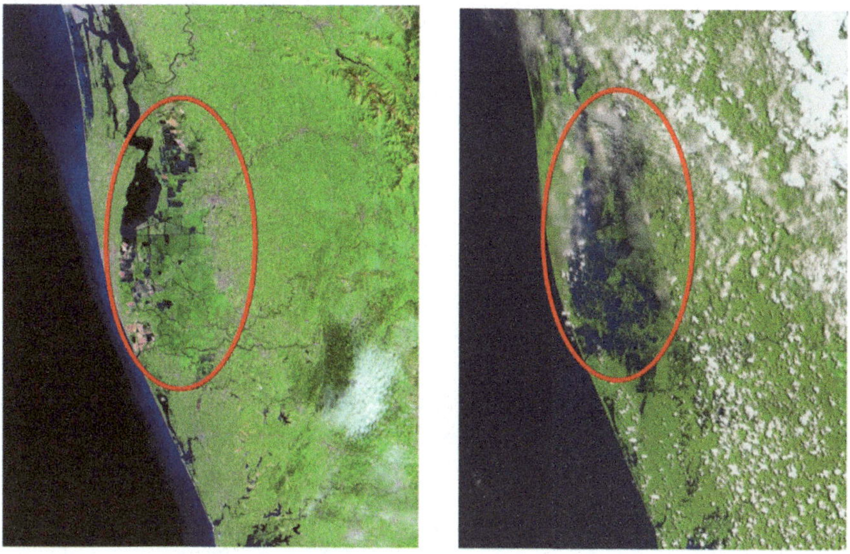

Fig. 7 Aerial view of Kochi region on August 1st 2018 (left) and August 14th 2018 (right) [60]

Table 1 Month-wise rainfall and percentage departure from normal [15]

Period	Normal rainfall (mm)	Actual rainfall (mm)	Departure from normal (%)
June 2018	649.8	749.6	15
July 2018	726.1	857.4	18
1–19 August 2018	287.6	758.6	164
Total	1649.5	2346.6	42

The storm of 15–17 August 2018 was spread over the entire Kerala with eye-centred at *Peermade*, a place between *Periyar* and *Pamba* sub-basins. The storm was so severe that the gates of 35 dams were opened to release the flood runoff. All 5 overflow gates of the *Idukki Dam* which is the largest arch dam in India were opened, for the first time in 26 years. Heavy rains in Wayanad and Idukki caused severe landslides and left the hilly districts isolated. On August 15, *Kochi* International Airport, *India's fourth busiest in terms of international traffic and the busiest in the State*, suspended all operations until August 26, following flooding of its runway. As per the reports in media, the flooding has affected hundreds of villages, destroyed several roads and thousands of homes have been damaged. The *Kerala State Disaster Management Authority* placed the state on 'red alert' as a result of the intense flooding. A number of water treatment plants were forced to cease pumping water, resulting in poor access to clean and potable water,

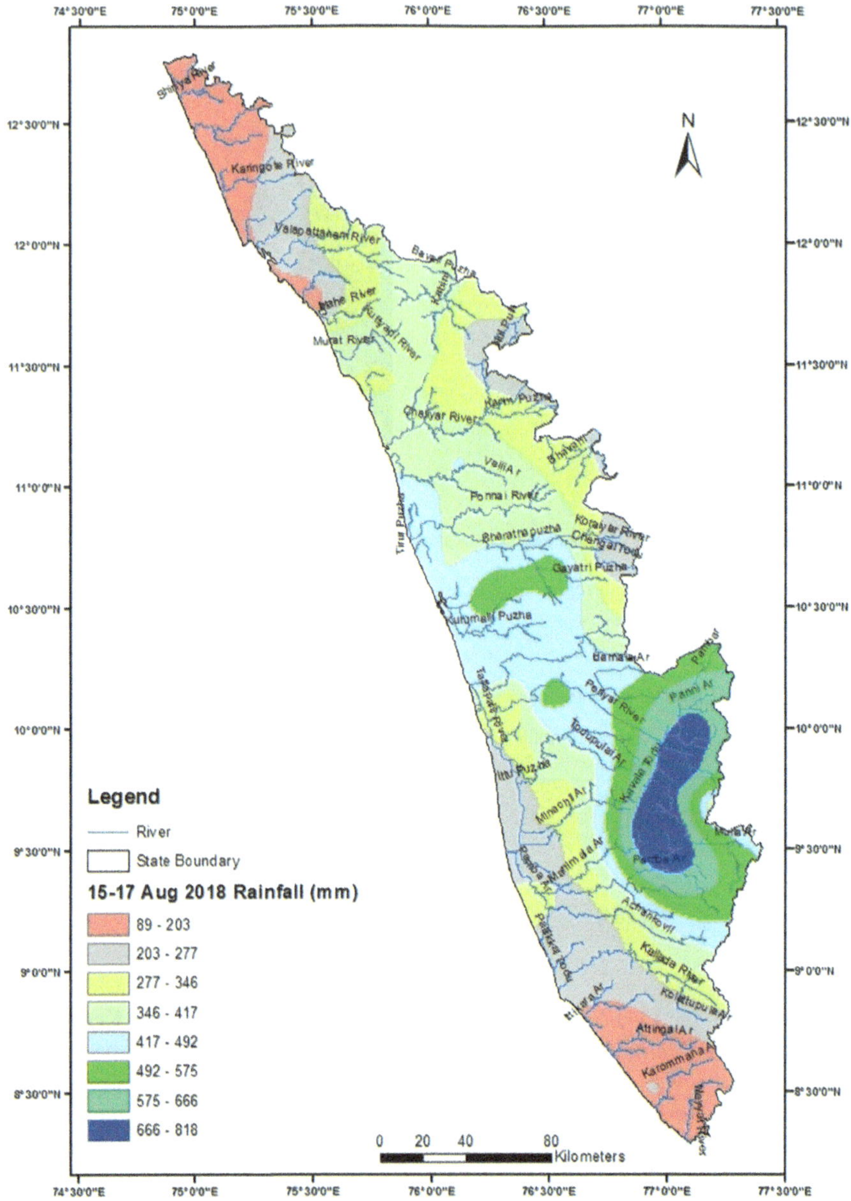

Fig. 8 3 days cumulative rainfall pattern of 15–17 August 2018 [15]

especially in northern districts of the state. A number of relief camps were opened to save the people from the vagaries of flood. The situation was regularly monitored by the State Government, Central Government, and National Crisis Management Committee which also coordinated the rescue and relief operations [15] (Fig. 9).

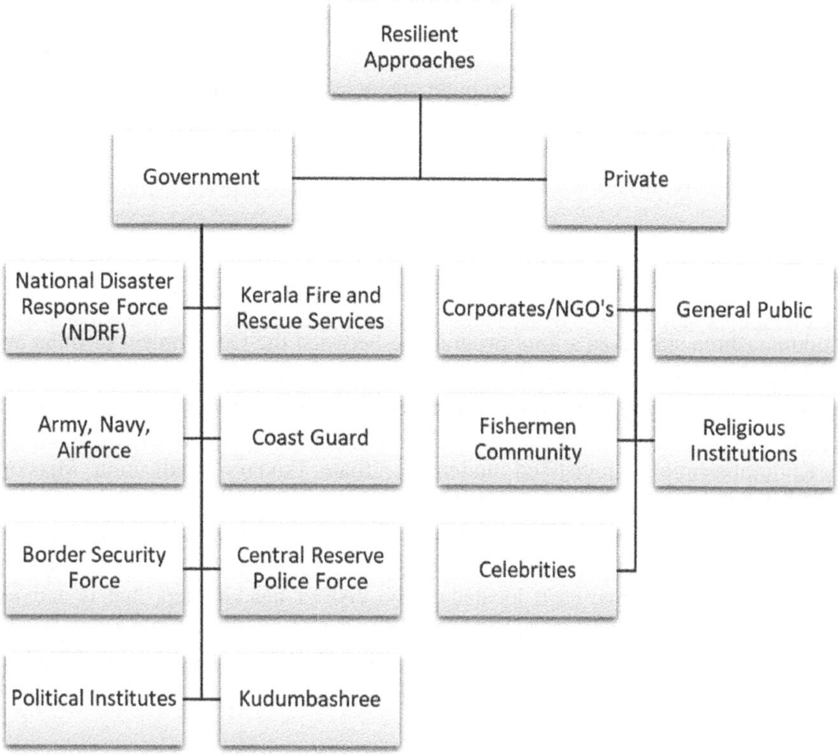

Fig. 9 Classification of Stakeholders involved in response and relief activities [15]

3 Response and Relief Approaches Adopted in Kerala

Government and private organisations were actively involved in the resilient approaches that led to a fast recovery of the state. It can be divided as shown in Table 2.

Table 2 NGO housing projects for Kerala flood 2018

NGO	No. of houses	Source
Co operative Department	1500	[40]
Muslim Jamaath	1000	[18]
Peoples Foundation	500	[34]
Act On	300	[4]
Joy Allukas	250	[50]
Muthoot Group	200	[29]

3.1 Government Organisations

Support from all involved government agencies was vital for resilience, since Kerala has not experienced a similar scale of disaster in nearly a century, the level of support required was unknown and the only way possible was to provide 'as much as support' the government could provide in every possible manner.

3.1.1 Kudumbashree

Kudumbashree started as a joint programme between the Government of India and NABARD, *National Bank for Agriculture and Rural Development,* focused on the upliftment of the economically weaker section of the community for women exclusively.

Kudumbashree is registered under the 'State Poverty Eradication Mission' (SPEM), a society registered under the Travancore Kochi Literary, Scientific and Charitable Societies Act 1955. It is one of the largest women-run organisations in the world and has a State Mission Office located at *Thiruvananthapuram* and 14 District Mission Teams, each located at the district headquarters that is actively involved in local community services across the state [39]. Apart from regular community services, the *Kudumbashree* was actively involved in the relief and rehabilitation works related to the flood relief operations. Some of the notable areas of involvement include:

- Support to cleaning more than 0.1 million houses and public offices
- Counselling support to more than 10,000 families
- Collection and distribution of relief materials from possible sources
- Support to volunteers in various flood relief activities
- Housing of flood victims in *Kudumbashree* exclusive houses as temporary shelters
- Cleaning of public properties
- Support to local self-government for all activities related to flood relief
- Support to local level coordination.

3.1.2 Political Parties

The flood situation was very grieving. Rather than blaming the ruling party for lack of preparedness for the disaster, *both the ruling and opposition parties,* joined hands in providing relief activities within the state. This helped in boosting quick state and national support to necessitate law and order in support of the relief activities [46]

3.2 *Private Organisations*

More than the governmental support, it was the private organisations that were involved as first-hand responders. The swift and timely support received helped save hundreds of lives and prevented cumulative effects of the disaster.

3.2.1 Corporate/NGO's

NGOs have been very active in providing services in various parts of need. Some of the housing projects for the flood victims by the NGOs are as follows:

3.2.2 Fishermen Community

The fishermen community played one of the most important roles in Kerala's resilience. Being first responders, they were able to rescue at least 65,000 lives [57]. A total of 4537 fishermen had risked their lives on 669 boats with very basic amenities to support the rescued [57]. However, timely intervention helped the stranded reach relief camps and get necessary first aid and other amenities. The eagerness of the fishermen community needs to be appreciated even when economically they might be in a very critical state. Community bonding is seen from such actions of the society (Fig. 10). Being able to navigate through water-borne areas with ease, they were the best help in terms of local support and knowledge of routes and people. Many rescuers claimed the boats of NDRF were too small and rescue was majorly due to fishermen's boats.

3.2.3 General Public

Support from common people was crucial for Kerala's resilience. Some of them being:

- *Setting up of Relief camps*
  Over 1 million people were displaced and stayed in relief camps across 3274 camps in the flood-hit districts [47, 48]. Many of these camps were set-up and run by the common public in safer locations such as schools, hospitals, playgrounds, markets, stadiums, religious places of worship *without differentiation* or places that were free from the flood. Anyone could donate food and other basic amenities. These supports were bought from different places to these relief camps for distribution, especially medicines and first aid. People were asked to stay in these camps until rehabilitation was complete. From the wholehearted support received, food supply was later abundant in most camps. After providing for the camp mates, this excess food was then given off to the poor and economically weaker section of the society.

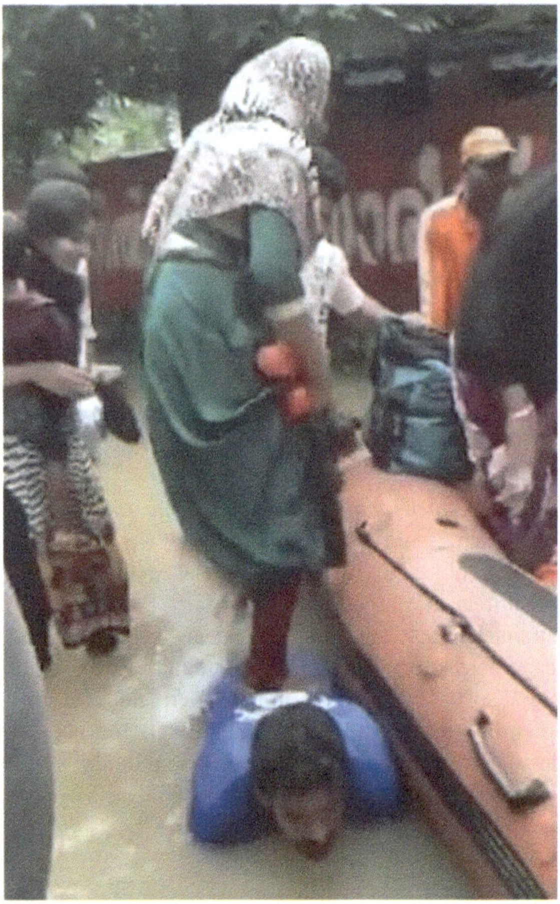

Fig. 10 A fisherman helping a lady board a rescue boat [62]

- *Local non-profit unrelated organisations*
  Apart from the fishermen community, other local non-profit organisations and social clubs too joined in the relief work activities by providing the resources that were available to them. For example, the *Off-Road Riders Club* in *Idukki* volunteered themselves with the help of their off-road vehicles to navigate through rough terrain inaccessible by normal vehicles to engage themselves in the rescue activities [47, 48]. Many of these vehicles are modified not in tune with the traffic regulations of the country (Fig. 11). However, at times of crisis, it was these modifications which helped save those extra lives. This also brings the question of amendments that needs to be made in the laws which are at odds with such requirements.
- *Social Media Influence*
  A lot of emergency information's such as road blocks, disrupted paths, availability of services at relief camps and even request for donations were widely

Fig. 11 Modified off-road vehicles being used by police for relief work [63]

propagated through multiple social media platforms such as *Facebook, WhatsApp and Twitter* [45]. The government made announcements were also spread by the people through these sources (Fig. 12). This helped people in remote areas to inform the authorities, pinpoint their stranded locations [6].

- *Transforming Spaces*
  Due to the flood condition in many areas, accessibility to better grounds for emergency treatments after rescue would sometimes be time-consuming. At such times, lives may be lost if not adhered immediately. This is when the available resources need to be utilised judiciously. Several religious institutes opened up their spaces for people of any community to rest, have food or even request for help that were within their jurisdiction. A mosque in *Malappuram* offered their mosque's prayer hall to be used as an autopsy room when time were against the odds [19].
- *Free Services*
  Due to flood, damage to property was inevitable. Completely and partially destroyed houses and loss of personal belongings affected the mental condition of any individual. At this time, many people tried to do whatever they can, to support their fellow citizens. Many free service centres for damaged vehicles in floods have been started by both two-wheeler manufacturing brands and private groups taking support from the local work force. This also helped reduce the high demand on authorised service centres post-disaster. Most people were able to get their vehicles running in a matter of days [49].

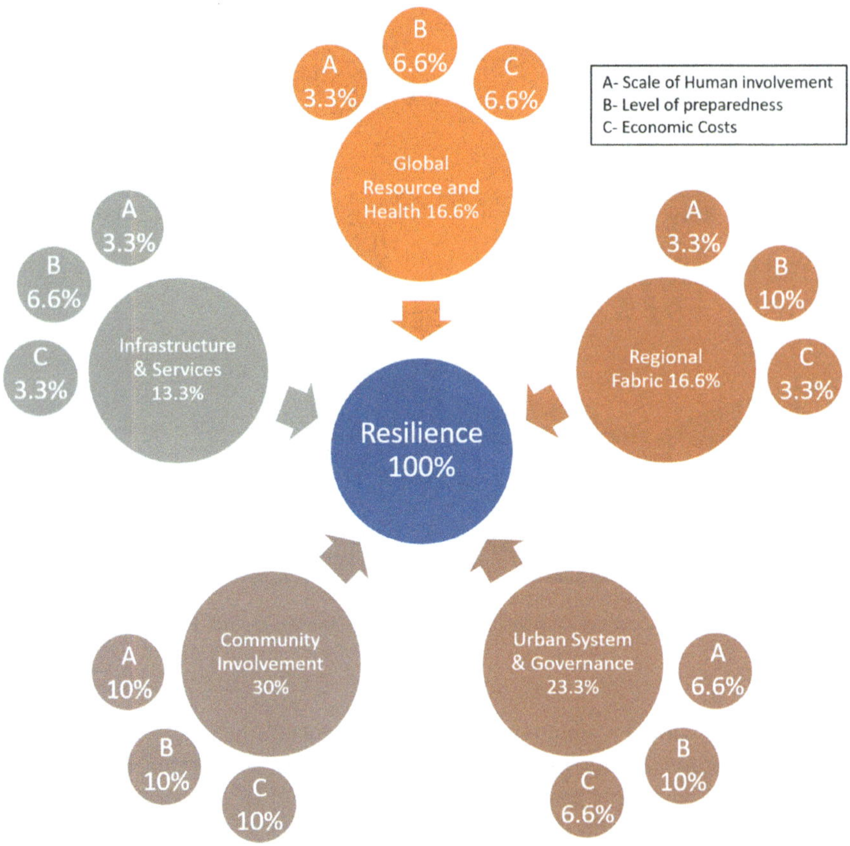

Fig. 12 Breakup of factor weightages contributing to urban resilience for calculating index for a given city

- *Celebrity Participation*
  Adding to the local support as first responders, the number of people and enthusiasm showcased further encouraged public participation in relief activities. Even celebrities such as movie actors were actively involved in the relief activities for days giving physical support as well as sharing vital information in the form of photos and video on social media platforms [44]. Public behaviour can be tuned to support a cause with the help of clarifications by known experts or well-known individuals' states [2] with the help of mass communication. The floods of 2019 also saw the return of these individuals to support the affected again in flood struck regions [32] (Table 3).

Table 3 Factors of resilience at different stages

Scale	Factors of relevance	References
1. Global	i. Environment protection	[56]
	ii. Resource inventory and utilisation	[11, 25]
	iii. Global population health	[5]
2. Regional	i. Economic fabric	[7]
	ii. Regional resource drift	[51]
	iii. Regional resource carrying capacity	[16]
3. Urban	i. Governance	[54]
	ii. Urban system	[3, 8]
	iii. Urban security	[38]
4. Community	i. Community needs	[27]
	ii. Neighbourhoods	[14]
	iii. Community management	[33]
5. Service	i. Infrastructure	[12]
	ii. Buildings	[42]
	iii. Transportation networks	[43]

4 Levels of Resilience

Resilience is found in all stages, namely global, regional, urban, community and service levels. At each level, the demands facility required and level of intervention required is different. All-in-all for a city to be truly resilient, every stage needs to be addressed with utmost priority as they are all dependent factors.

Many of the above studies prefer multiple connections between the scales of resilience. The interconnectivity is yet another area that needs to be explored. Community resilience is the least cited subject among most studies, but the studies that do suggest them accentuate its impact it makes on resilience [33] highlights the influence the fishing community has had on the city's resilience (Philippines) during times of crisis, so had been the case of Kerala as well during its flood in 2018.

Community involvement changed the complete perspective of urban resilience in Kerala during and after the event of the flood. Rather than the built-forms, it was the *human effort* that helped achieve Kerala its true resilience. Compared to any similar scale of disaster around the globe, Kerala was able to be truly resilient in just a matter of months. 'The people and the state worked in tandem, a rare sight for an Indian State in crisis', quoted by [37] of the Observer Research Foundation states that the state achieved its previous condition seamlessly and much quicker than usual when compared anywhere else of a similar magnitude [37]. Also points out the fact that managing the disaster efficiently and clinically was comparable to that of most developed nations. The youth of the state were too actively involved in

Table 4 Weights assigned to different levels of resilience

Levels of resilience	Scale of human involvement (1–2–3)	Level of preparedness involved (1–2–3)	Economic costs involved (Post-disaster) (1–2–3)	Total Points (3–9)
1. Global resources and health	1 Strength of international relations	2 An endemic outbreak is critical	2 Global support available but keeps debt	5
2. Regional fabric	1 Depends on political relations	3 Regional maps and trend studies	1 Depends again on political relations	5
3. Urban system and governance	2 Includes public participation	3 Zonal and hazard maps	2 Need support usually	7
4. Community involvement	3 Can be actively involved both during pre and post disaster	3 Floods of 2019 showed higher levels of preparedness	3 Close to nil since its usually donations	9
5. Infrastructure and services	1 Public property	2 Unpredictable events reduces preparedness	1 Resilient structures are costlier	4

running relief camps and mobilise resources [22], states how they can serve as templates to the rest of the country on how to unite during times of crises.

From the above factors of resilience studies and the relief approaches adopted in Kerala, a chart has been prepared including the factors of involvement, preparedness level and costs involved in order to obtain a 9-point scale with 3 points under each category (higher the number, better the resilience). 1 corresponding to low score, 2 for medium and 3 for high; 0 score has not been assigned since all are factors of resilience as adopted from above-mentioned studies.

From Table 4, it is seen that resilience through community involvement is the highest factor of resilience across various factors of involvement. The costs involved are the lowest, external dependencies are the lowest and the time of interaction is the fastest. The situation in Kerala helped achieve a new fabric in resilience that was previously overlooked. A positive community relationship is necessary however to achieve this state. Cities need to majorly focus on how to bring their communities together for a myriad of factors and community resilience is only one that adds to it.

5 Discussions

The state of Kerala was able to manage the flood disaster efficiently despite the lack of preparedness with *Community participation* being the key factor especially in its case. The overwhelming support from different groups of people and organisation helped reach Kerala to its stable state within a few months post disaster. Some of the takeaways and recommendations to increase resilience are:

- Encourage the *ethnic and economic integration* of different neighbourhoods.
- *Mix housing patterns and create common public spaces* for increased interaction.
- *Blend different houses of worship of all faiths* from the grassroots of city planning.
- *Plan for the pandemic by providing for adequate municipal and private services*, even in the event of the disaster of large proportions.
- Be prepared with *stock of extra amenities available at all times*, especially potable water.
- *Create special cells for constant monitoring and also vigilant quick response teams* dedicated to different areas. Create a *zonal rebuilding map and integrate with special cells.*
- *Ensure basic services especially for the poor neighbourhoods*, since they are the keystone for the population of any metropolitan region, that is, establish standards for housing unit loss per unit population per year.
- *Ensure buildings reconstructed are using disaster resilient techniques* and away from flood plains and slopes.
- *Create public awareness and prepare people to respond in time of crisis,* both in affected and unaffected areas.
- Restoration of irrigation-cum-drainage systems.
- Prepare an update *Hazard Vulnerability map for Kerala* and also establish methods of evacuation or resistance at times of Risk.
- Establish *Resilience Efficiency Index* to establish level of preparedness and assistance.

Any disaster is an opportunity to establish a robust human rights-based approach at all phases of recovery cycle, based on the principles of *non-discrimination, participation and 'leaving no one behind'* as per Agenda 2030 [57]. On an alternate dimension, frequent occurrences of disasters in the same place have also raised the eyebrows of sceptics claiming this to a result of conspiracy theory. Occurrences of flash floods and unpredictable weather conditions further heighten this sceptical thinking. The chances for a cloud-seeding event due to disputes with neighbouring states/countries cannot be out-ruled either [1].

6 Conclusion

Encouragement of public participation is the key factor contributing to a city's resilience. *Feelings of oneness,* keeping aside personal outlooks such as political, religious or caste-based classification are factors the government should focus to implement at times of crisis for a faster resilience. The potential for *Green Infrastructure* is another aspect that is gaining rapid attention which has the potential to reduce damages from natural disaster especially after the catastrophic events. *Preservation of natural ecosystem and flood plains and integrating them into the urban fabric* will increase community resilience while providing social, economic and environmental benefits. The factors of resilience decoded from the studies of Kerala further clarify the importance of community involvement that needs to be made a standard for resilience planning for any region or city. The outcome will be a healthier community with higher resilience to future events of disasters. Further, establish *Resilience Efficiency Index* to establish level of preparedness and assistance even to budgeting and fund allocation at governmental level.

References

1. Abraham B (2018) From Lord Ayyappa to beef and dogs, here's everything that's been blamed for Kerala floods. Available at https://www.indiatimes.com/news/india/from-lord-ayyappa-to-beef-and-dogs-here-s-everything-that-s-been-blamed-for-kerala-floods-351623.html
2. Abroms LC, Maibach EW (2008) The effectiveness of mass communication to change public behavior. Ann Rev 219–234
3. Aerts JC, Botzen WW, Emanuel K, Lin N, De Moel H, Michel-Kerjan EO (2014) Evaluating flood resilience strategies for coastal megacities. Science 473–475
4. Anon (2018) Information and Public Relations Department. Available at http://prd.kerala.gov.in/ml/node/23294
5. Asprone, D, Manfredi G (2015) Linking disaster resilience and urban sustainability: A glocal approach for future cities. Disasters S96–S111
6. Aswathi SB, Dinesh Babu S, Harikrishnan D (2019) Impact of social media in dissemination of information during a disaster—a case study on Kerala floods 2018. Int J Inno Technol Exploring Eng 283–286
7. Barata-Salgueiro T, Erkip F (2014) Retail planning and urban resilience—an introduction to the special issue. Cities 107–111
8. Barthel S, Parker J, Ernstson H (2015) Food and green space in cities: a resilience lens on gardens and urban environmental movements. Urban Stud 52(7):1321–1338
9. Béné C, Mehta L, McGranahan G, Cannon T, Gupte J, Tanner T (2017) Resilience as a policy narrative: potentials and limits in the context of urban planning. Clim Dev 116–133
10. Budruk M, Thomas H, Tyrrell T (2009) Urban green spaces: a study of place attachment and environmental attitudes in India. Soc Nat Resour 824–839
11. Campanella T (2006) Urban resilience and the recovery of New Orleans. J Am Plann Asso 141–146

12. Chang SE, McDaniels T, Fox J, Dhariwal R, Longstaff H (2014) Toward disaster resilient cities: characterizing resilience of infrastructure systems with expert judgments. Risk Anal 416–434
13. Chattopadhyay S, Franke RW (2006) Striving for sustainability: environmental stress and democratic initiatives in Kerala. Striving for sustainability. Concept Publishing Company, New Delhi, pp 110–112
14. Chelleri L, Schuetze T, Salvati L (2015) Integrating resilience with urban sustainability in neglected neighborhoods: challenges and opportunities of transitioning to decentralized water management in Mexico City. Habitat Int 122–130
15. Directorate H (2018) KERALA FLOODS OF AUGUST 2018. New Delhi: Central Water Commission
16. Davoudi S (2009) Scalar tensions in the governance of waste: The resilience of state spatial Keynesianism. J Environ Plann Manage 52(2):137–156
17. Department I-IM (2019) India Meteorological Department. Available at http://www.imd.gov.in/pages/alert_view.php?ff=20190812_al_1019
18. Hindu T (2018) Jamaath plans 1000 houses for flood victims. Available at https://www.thehindu.com/news/cities/kozhikode/jamaath-plans-1000-houses-for-flood-victims/article24897130.ece
19. India Today (2019) Kerala floods: Mosque offers prayer hall to be used as mortuary. Available at https://www.indiatoday.in/india/story/kerala-floods-mosque-offers-prayer-hall-to-be-used-as-mortuary-1581086-2019-08-15
20. Kerala State Disaster Management (2018) PDNA—Post disaster needs assessment. Kerala State Government, Thiruvananthapuram
21. King AD (1976) Colonial urban development-culture, social power and environment. Routledge, London
22. Kochukudy A (2018) Anatomy of a flood: how Kerala withstood a calamity. Available at https://thewire.in/environment/kerala-floods-pinarayi-vijayan-idukki-chalakudy
23. Kumpfer KL (2002) The resilience framework. In: The resilience framework, pp 179–224
24. Laituri M, Kodrich K (2008) On line disaster response community: people as sensors of high magnitude disasters using internet GIS. Sensors 3037–3055
25. Leichenko R (2011) Climate change and urban resilience. In: Current opinion in environmental sustainability. Elsevier, London, pp 164–168
26. Mark Pelling SB (2013) Megacities and the coast-risk, resilience and transformation. Routledge, London
27. Mehmood A (2016) Of resilient places: planning for urban resilience. Euro Plann Stud 407–419
28. Mission S (2018) Suchitwa mission damage assessment report. Government of India, New Delhi
29. News Time Network (2018) News time network. Available at https://www.newstimenetwork.com/the-muthoot-group.html
30. Office of the Registrar General & Census Commissioner, I (2011) Census of India 2011. Ministry of Home Affairs, Government of India, New Delhi
31. Oliver-Smith A (1996) Anthropological research on hazards and disasters. Ann Rev Anthropol 303–328
32. Onmanorama Staff (2019) Tovino Thomas wins hearts as he visits relief camps. Available at https://english.manoramaonline.com/entertainment/entertainment-news/2019/08/12/tovino-thomas-flood-relief-activities-kerala-social-media.html
33. Pauwelussen A (2016) Community as network: exploring a relational approach to social resilience in coastal Indonesia. Marit Stud 15(2)
34. People's Foundation (2018) Flood relief rehabilitation project distribution. Available at https://peoplesfoundation.org/event/wordcamp-paris-5/
35. Prasad R (2018) A rare confluence of events led to flooding in Kerala. IIT Gandhinagar, Gandhinagar

36. Radbound U (2019) Subnational human development index. Available at https://globaldatalab.org/shdi/shdi/
37. Sen S (2018) Improved disaster management saves Kerala, despite lack of preparedness. Available at https://www.orfonline.org/expert-speak/43901-improved-disaster-management-saves-kerala-despite-lack-of-preparedness/
38. Server OB (1996) Corruption: a major problem for urban management: some evidence from Indonesia. Habitat Int 23–41
39. State Poverty Eradication Mission, G. o. K (2019) Organisational structure. Available at http://www.kudumbashree.org/pages/8
40. Sumeesh (2018) Relief for Kerala flood victims (in local language). Available at http://malayalam.webdunia.com/article/kerala-news-in-malayalam/pinarayi-vijayan-118090100033_1.html
41. Suryanarayana MH, Agrawal A, Prabhu KS (2011) Inequality-adjusted human development index for India's states. UNDP India, New Delhi
42. Takewaki I, Fujita K, Yamamoto K, Takabatake H (2011) Smart passive damper control for greater building earthquake resilience in sustainable cities. Sustain Cities Soc 3–15
43. Testa AC, Furtado MN, Alipour A (2015) Resilience of coastal transportation networks faced with extreme climatic events. Transp Res Record 29–36
44. The Better India (2018) Kerala floods: Malayalam actor sets example, opens house to victims who need shelter!. Available at https://www.thebetterindia.com/156440/tovino-thomas-actor-kerala-floods-news/
45. The Caravan (2018) How social media was vital to rescue efforts during the Kerala floods. Available at https://caravanmagazine.in/vantage/how-social-media-was-vital-to-rescue-efforts-during-the-kerala-floods
46. The Hindu-Business Line (2018) Kerala political parties unite in call for 'national calamity' tag. Available at https://www.thehindubusinessline.com/news/kerala-floods-kerala-political-parties-unite-in-call-for-national-calamity-tag/article24738418.ece
47. The Times of India (2018a) Kerala floods: over 1 million in relief camps, focus on reha …. Available at https://timesofindia.indiatimes.com/india/kerala-floods-over-1-million-in-relief-camps-focus-on-rehabilitation/articleshow/65478086.cms
48. The Times of India (2018b) Off-road riders turned saviours in flooded Idukki. Available at https://timesofindia.indiatimes.com/city/kochi/off-road-riders-turned-saviours-in-flooded-idukki/articleshow/65570871.cms
49. The Times of India-City (2018) Kerala: flood-hit bikers get a helping hand. Available at https://timesofindia.indiatimes.com/city/kochi/kerala-flood-hit-bikers-get-a-helping-hand/articleshow/65641796.cms
50. TOI (2018) Available at https://timesofindia.indiatimes.com/city/kochi/joyalukkas-group-extends-support-to-rebuild-kerala/articleshow/65852909.cms
51. Toubin M, Laganier R, Diab Y, Serre D (2015) Improving the conditions for urban resilience through collaborative learning of Parisian urban services. J Urban Plann Dev 141(4)
52. Trogal K, Bauman I, Lawrence R, Petrescu D (2018) Architecture and resilience: interdisciplinary dialogues. Routledge, New York
53. United Nations Development Programme (2018) Kerala—Post disaster needs assessment-floods and landslides—August 2018. Government of Kerala, Thiruvananthapuram
54. Wagenaar H, Wilkinson C (2015) Enacting resilience: a performative account of governing for urban resilience. Urban Stud 1265–1284
55. Wallace D, Wallace R (2008) Urban systems during disasters: factors for resilience. JSTOR
56. Woolhouse MEJ, Rambaut A, Kellam P (2015) Lessons from Ebola: improving infectious disease surveillance to inform outbreak management. Sci Transl Med 307rv5
57. World Bank (2018) Kerala Post disaster needs assessment- Floods and Landslides August 2018. United Nations Development Programme, Thiruvananthapuram
58. Centre KS (2016) Kerala state disaster management plan 2016. Thiruvananthapuram: Kerala State Disaster Management Authority

59. India Times (2018, August 11). The Kerala Deluge Brings Back Memories Of 'The Great Flood Of 99' Which Almost Wiped Out Munnar. Retrieved from India Times: https://www.indiatimes.com/news/india/the-kerala-deluge-brings-back-memories-of-the-great-flood-of-99-which-almost-wiped-out-munnar-351058.html
60. Observatory NE (2018, August 25) Before and After the Kerala Floods. Retrieved from NASA Earth Observatory: https://earthobservatory.nasa.gov/images/92669/before-and-after-the-kerala-floods
61. Bhuvan (2019, 31 August) Digital Elevation Model (DEM) generated from Cartosat-1 Satellite data for India. Retrieved from data.gov.in: https://data.gov.in/catalog/digital-elevation-model-dem-generated-cartosat-1-satellite-data-india?filters%5Bfield_catalog_reference%5D=164389&format=json&offset=0&limit=6&sort%5Bcreated%5D=desc
62. ABP Live (2018, August 19) Kerala floods: Meet the unsung heroes of rescue mission in rain-ravaged state. Retrieved from ABP Live: https://www.abplive.in/india-news/kerala-floods-latest-news-the-unsung-heroes-of-rescue-operations-in-rain-ravaged-state-744975
63. The Financial Express (2019, January 19) SUV owners who rescued people in Kerela floods upset with ban on modified cars. Retrieved from The Financial Express: https://www.financialexpress.com/auto/car-news/suv-owners-who-rescued-people-in-kerela-floods-upset-with-ban-on-modified-cars/1458293/

Reinforcing the Long Forgotten Southern Frontier of Madras

G. Bhuvaneshwari and Girishma Kongara

Abstract Stretching for about 22 kms inside the Chennai City, the Adayar River is currently a sewage channel that is running across the city. Dating back in the history around nineteenth century, Adayar was a non-perennial river which was used for seasonal recreation by the British. Madras being home to three main rivers Kosasthalaiyar, Cooum, and Adayar was home to the first rowing center in Madras. Madras Boat Club which is located in the banks of Adayar is the regulatory body to conduct boating activities. The boating activities slowly shifted from the Ennore creek to long tank and then to the Adayar. Due to the rapid urbanization and increasing demand for infrastructure on one hand and land unavailability on the other, the acquisition of land on the river beds inside the city reached its peak. Connecting the sewage drains to the river, dumping of waste in and on the banks and multiple other factors including e-waste and bio-waste deposits in the water has led to the deterioration of the river's ecology. This research will investigate the bathymetry of the river, the behavior of the residents of Saidapet settlement and the functioning of Guindy Industrial Estate, which lie along the banks of Adayar from Ekkaduthangal bridge and Marmalong bridge in the heart of Chennai City, and come up with a sample model which will focus on indigenous system of water quality improvement that will aid in the natural process of runoff and infiltration and also complement to reduce the impact of flooding, thus bracing Adayar.

Keywords Madras · Adayar · Ecology · Indigenous method · Saidapet · Constructed wetlands · Riparian buffer · Biodiversity

---

G. Bhuvaneshwari (✉) · G. Kongara
School of Architecture and Planning, Anna University, Chennai, Tamil Nadu, India
e-mail: bg98.ar@gmail.com

G. Kongara
e-mail: girishma.kongara02@gmail.com

K. Thirumaran et al. (eds.), *Sustainable Urban Architecture*,
Lecture Notes in Civil Engineering 114,
https://doi.org/10.1007/978-981-15-9585-1_9

1 Introduction

The backbone of an urban fabric is its natural resources and the infrastructural facilities that offer the citizens a good quality of living and sustenance. Natural resources play an important role for the sustenance of living and non-living things on earth. They imbibe an interaction between them from which human being's benefit. People tend to use the resources to survive and developing countries consume more amount of resources. These natural resources tie hand in hand with climate change. A global phenomenon which is causing a lot of confusion and threats in the world, climate change, in such a case, rivers and other water bodies play a vital role in the development of a city and also directly affect the climate change. Climate change has the potential to form an imbalance to the base of a civilization which the world is witnessing currently. Major changes that we are facing include the shift in seasons, availability of water in summer season, the extreme cases of flood or drought. This has created a situation where the long thought calculative design decisions are questioned and have created a disaster due to these unforeseen changes. In one influential academic paper, scientists proposed that "stationarity is dead," which means that over time the expectation of a relatively stable climate and narrow range of precipitation patterns are no longer assured.

In such a scenario, understanding and responding to this issue are vital. This research began with analysis of the stretch of Adayar within the urban fabric of Chennai.

2 Methodology

The study began with a basic understanding of the city's important resource Adayar, which has significance of past and evidence of present. Based on this, an extent was chosen within the river's boundary in the city (see Fig. 1). A clear difference between the chosen extent and rest of the river is that there was water in the river bed yearlong whether or not there was rain. This extent of the river has a distinct character which is the stark difference (see Fig. 2) in the land use pattern on its either sides, which was the Guindy Industrial Estate and the Saidapet/Jafferkhanpet settlement. These facts laid the foundation for the analysis and the study. The major content of our study was finding data with respect to the chosen extent of the Adayar, the behavior of the local residents in relation to the river, and the Adayar's bathymetry.

Following aspects were looked into and analyzed to arrive at a conclusion

- History of Adayar—the river.
- History of the settlements and Guindy Industrial Estate.
- Hydrology of Adayar.
- Ecology of Adayar.
- Edge conditions of the river.
- Influential factors for the current state of the river.

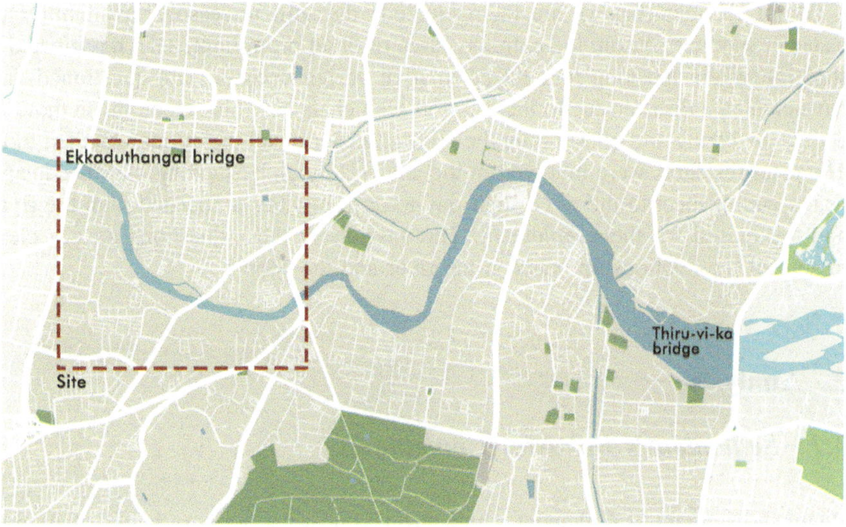

Fig. 1 Chosen extent of Adayar with respect to its flow through the major part of the city

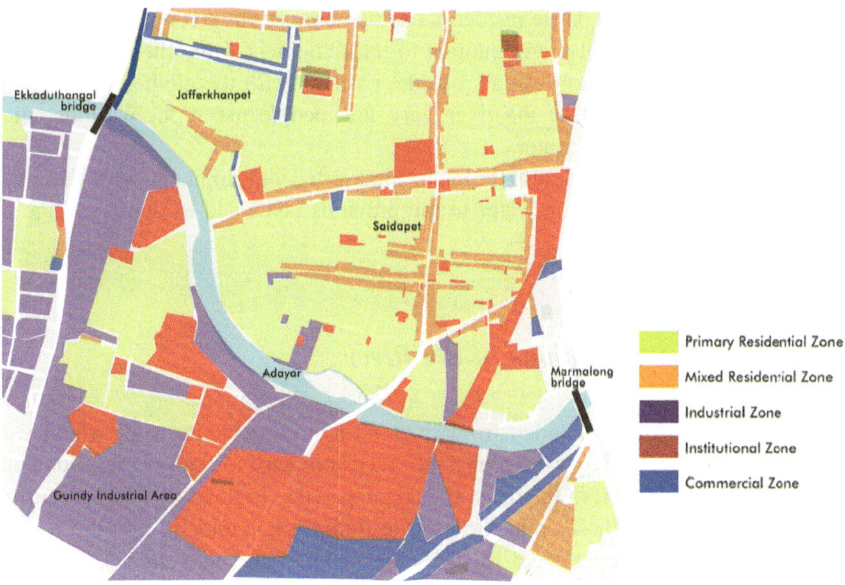

Fig. 2 Land use of the urban fabric within the extent of study

Major changes that we are facing include the shift in seasons, availability of water in summer season, the extreme cases of flood or drought. This has created a situation where the long thought calculative design decisions are questioned and have created a disaster due to these unforeseen changes. This helped the authors in deeply analyzing and questioning the current scenario and the reasons that led to it. The local residents were involved in the analysis by understanding their needs and wants with respect to the river, their day-to-day activities associated with the river and the reason for their settlement. The answers to these questions gave a clear picture of the relationship between the river and the people.

3 Analysis

3.1 Settlement's Behavior Contribution

Ragunandhapuram, a settlement which evolved near the northern banks of Adayar River, was evolved around three main temples—Karaneeshwarar temple, Perumal temple, and Murugan temple. The temples were the centric forces for the evolution of individual residential blocks. But through time the cities growth influenced Saidapet to evolve into a dense residential settlement and the river has always been only a backdrop making it more ignorant to the settlement. Only the lower income people which were the dhobi wala's use the river to wash the clothes but through time due to the pollution of the river very few people use it for washing after filtering and purifying the water.

The industrial estate has also been an introverted development with the river only been used for extracting water in the previous years and to dump sewage or industrial waste illegally (Fig. 3).

3.2 The Edge Conditions of the River

The slum settlement is secluded from river by a stone retaining wall which was constructed during MGR period along the river boundary to protect the settlement from flooding. Some part of the old wall is damaged due to flooding which occurred in the year 2015, and these spots are used to dump waste directly into the river.

The new wall is constructed on the other side of the river as a part of the government proposal, demarking the river boundary it was also constructed to protect the warehouses and the factories from theft as people can easily access the industrial estate through the river when it is dry. There are draining and dumping activities found even along this wall (Fig. 4).

Fig. 3 Ground map of the selected stretch with landmark

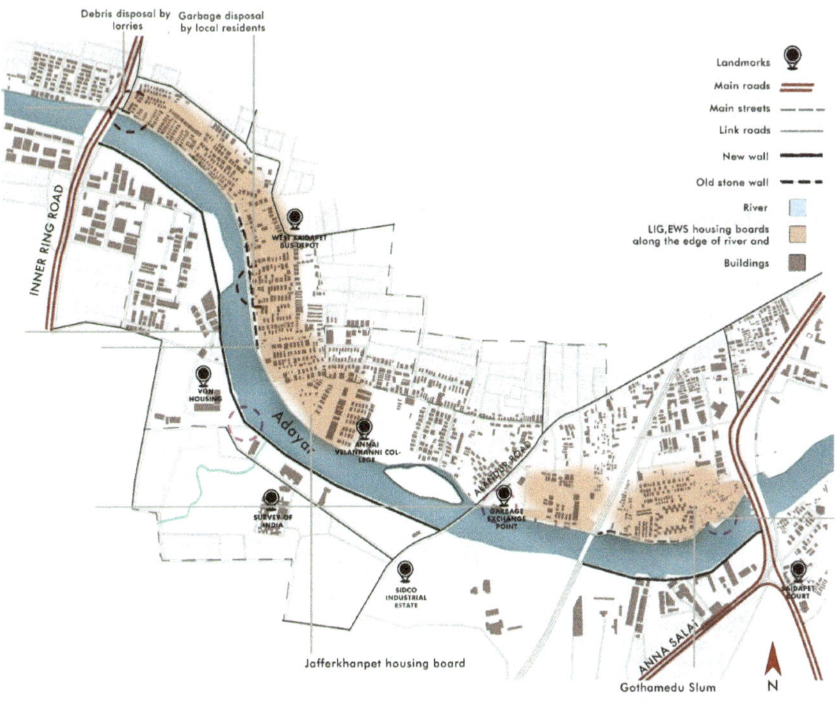

Fig. 4 Edge conditions

3.3 Pollution

3.3.1 Chemicals in the River

Biological oxygen demand—The total amount of oxygen used by microbes to breakdown the organic material in an aquatic ecosystem is called BOD. The values vary from 59 to 87 mg/L. The high values of BOD are due to increased inflow of organic matter as well as domestic sewage in the water body.

PH—On the whole, the pH of the water samples is under the considerable limit (range 5.5–9.0, according to general standards of Environmental [protection] rule, 1986).

Fluoride—The maximum permissible limit of fluorides is 2 mg/l as per the Environmental (protection) rule, 1986.

Electrical conductivity—The maximum permissible limit is about 1.4 mhos. Total suspended solids—The standard limit according to EPR, 1986 is 100 mg/L.

Chemical oxygen demand—The maximum permissible limit for COD is 250 mg/L.

Chlorine—On the whole, the chlorine content of the samples collected is within the limit 1 mg/L as specified by the EPR, 1986.

Total Kjeldhal Nitrogen—The standard value specified by EPR, 1986 is 100 mg/L. Turbidity—The general standards for turbidity as prescribed by WHO is 5NTU.

Dissolved oxygen—The permissible limit, according to WHO standards, ranges in between 6.5 and 8.5 mg/L.

From the Fig. 5, we can see that the water quality of Adayar River has deteriorated seriously and would continue if measures are not taken place to control the waste discharge into the river (Figs. 6, 7, 8, and 9).

3.3.2 Major Drainage Canals to the Adayar River

There are ten major storm water drains to the canal to carry excess surface flow to the river, but they have been exploited and have been illegally connected to the sewage drain systems of the nearby residences and been polluted. The nearby STPs also let a considerable amount of untreated water into the river (Figs. 10 and 11).

LOCATION	Ph	TEMP (^{o}C)	TURBIDITY	FLUORIDE	CHLORIDE	EC	TS mg/l	TSS mg/l	TDS mg/l	COD 1ML	COD 2 ML	COD 2.5 ML	TKN mg/l	CHLORINE	NITRITE mg/l	DO mg/l
Jafferkhanpet	7.30	26	26	3.0	436.04	1.76	118400	7840.46	110919.54	816	816	816	210	0.2	0.2	5.33
Saidapet	6.99	24	36	1.5	779.90	1.59	20300	16777.27	3522.27	272	272	272	210	0.2	0.5	5.00

Fig. 5 Chemical levels [1]

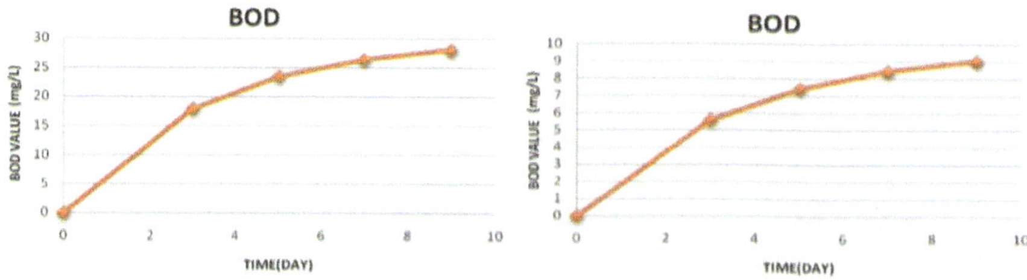

Jafferkhanpet y **Saidapet**

Fig. 6 BOD levels

Fig. 7 Panoramic view of the river

Fig. 8 Present conditions of the river

Fig. 9 Existing stone wall to protect the settlement from flooding

3.4 Rivers Topography Affecting Hydrology

Sections along the river have been taken to understand the topography of the river. The extent of encroachment is studied (Figs. 12 and 13).

While studying the longitudinal section (Fig. 6) along the river, it is evident that the gradient of the slope is not uniform. The undulating terrain form makes the water stagnant at many spots. And the high point acts as a barrier and restricts the solid waste from moving father toward the ocean, making this stretch the most polluted (Figs. 14 and 15).

3.5 Ecology

The entire stretch of Adayar River, due to its varying gradient and its proximity to the sea, has its merits and demerits. The part of the river closer to the sea, because of the tidal waves it washes out the impurities naturally, unlike the other areas which makes it sustain its ecological balance (Fig. 16).

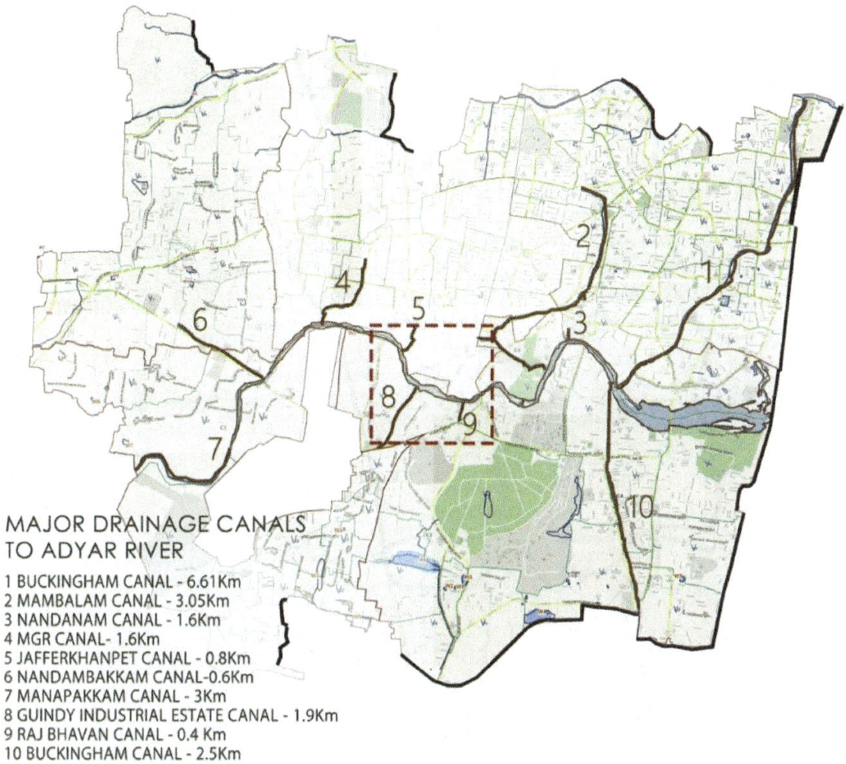

Fig. 10 Major canals draining into Adayar

The major contributors for Adayar's deteriorating ecology are A. Bio medical wastes from Thiruneermalai and MIOT Hospital (started around 2011) B. The development of MGR Nagar (1970s) whose sewage lines were directly connected to the river. The local residences' contribution in terms of garbage, sewage, and using it for sanitary purposes.

3.6 *Garbage Point*

There are seven major garbage transfer stations in Chennai City including Pudupet, Basin bridge, Elephant gate road, Oteri, Mylapore, Nungambakkam, and near Alandur Road in Saidapet. The garbage from every individual street is collected by Garbage trucks and dumped at pits in garbage transfer stations where it is stored for 24 h to reduce liquid content. Then, it is transferred to nearby dump yards (Perungudi or Kodungaiyur) by dump trucks. The Saidapet garbage transfer station was established 8 years ago which collects garbage from wards under ZONE 13 of

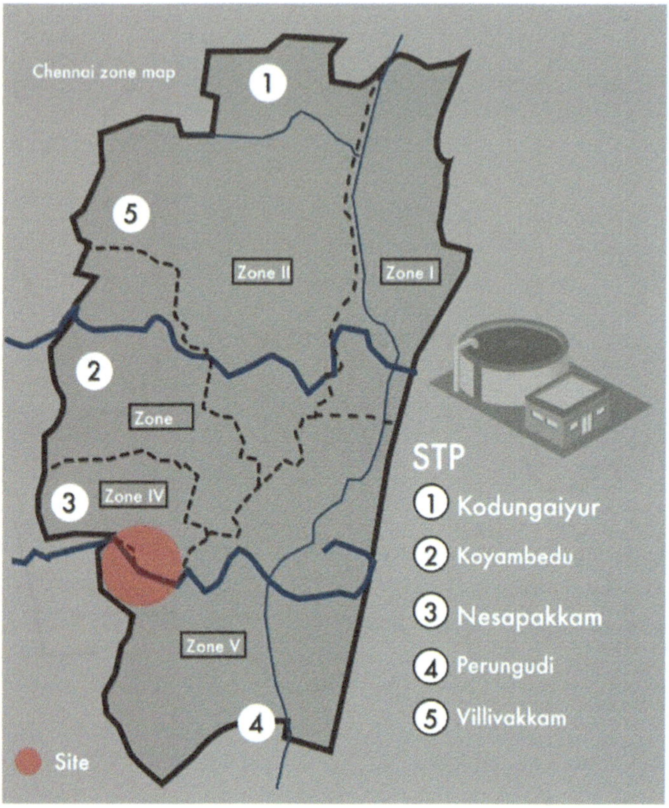

Fig. 11 STPs located in the city

Chennai City. There are 24 garbage trucks dumping waste from areas including Guindy West, Saidapet, Kotturpuram, Adayar, Besant Nagar, Velachery, Thiruvanmiyur, and Taramani. The people who live around the area and this garbage exchange point thrive out of this station by working as rag pickers, collectors and sellers from which they sustain their daily life. They are settled along the Alandur road and near the Alandur bridge that connects Saidapet and Guindy. Major percentage of these people belongs to the lower/working class. Their average income is ₹300–₹400 depending upon the quality of the waste and the recyclability character of the same. The vulnerability to risks of these workers is considerably high (Fig. 17).

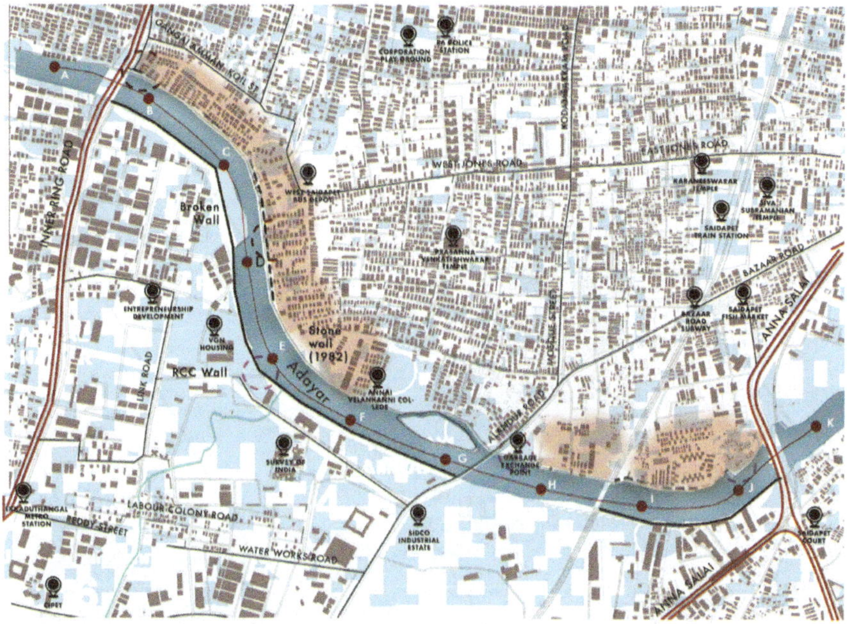

Fig. 12 Flood inundation

4 Proposal by State and Central Government

The main aims are to decongest Anna Salai, Inner ring road, Kathipara and Mount Poonamallee road and to revive the waterways along Adayar. The major schemes and projects focus on creating at Adayar Estuary and create an axis along the river to reach it. Additionally, ferry services/jetties to restart the waterways are proposed. Funds have been allotted for the initiation of these projects, and evidence of its progress is seen in places like YMCA college of physical education, Alandur bridge and near the estuary. Delineating the river boundary and creating a buffer of 10 meters on either side of the river to create a feasible site to implement the proposals. The proposal's initial plan of action will involve cleaning the river and organizing sewage disposal among settlements. The allied reasons for this project are to mitigate the flood's effect along the river plain. The Chennai River Restoration Trust monitors the actions of the other bodies and manages the eco-restoration project. Overall 555 crores of Indian rupees have been allotted, and around 17,000 families have to relocated. This proposal by the government aided a helping hand for collecting resources and also in recommendations proposed by the authors (Fig. 18).

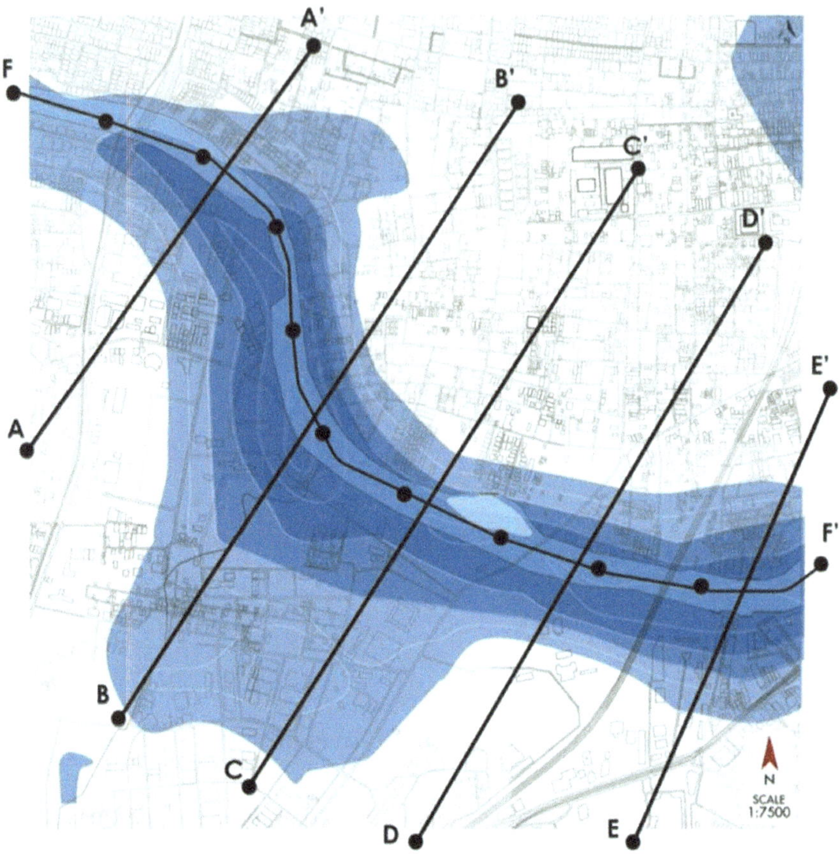

Fig. 13 Map showing the sectional lines cut along the river

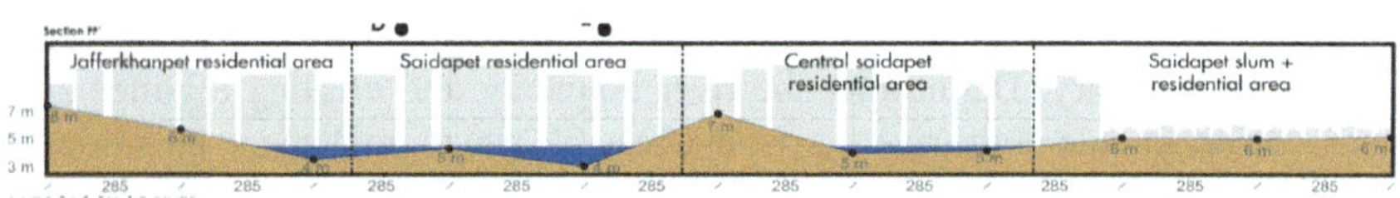

Fig. 14 Section FF'

5 Recommendations

The analysis of the chosen extent of study and the findings, the following recommendations/proposals are given for improving the situation of the river, contribute to the sustenance of the city, and on the whole to support the urban ecology. These recommendations were derived also based on understanding the flood inundation of 2015 in Chennai and the critical contour of the river bed.

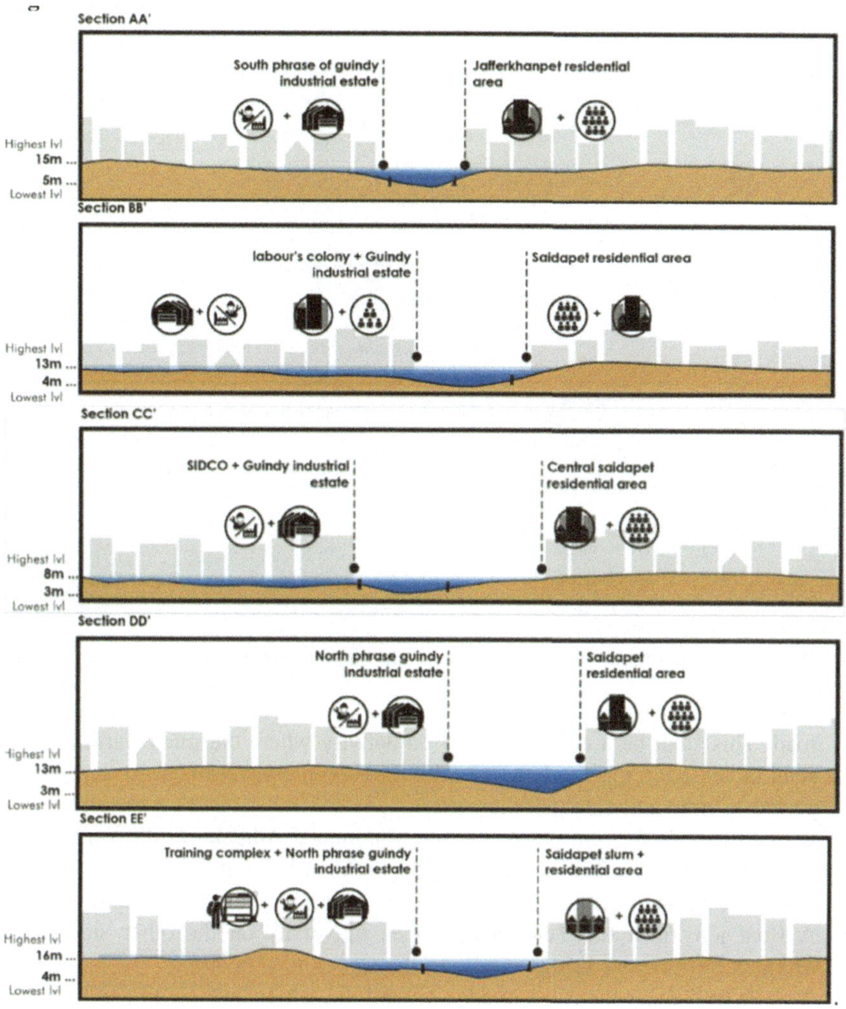

Fig. 15 Section across the river

- Riparian belt along the critical contour to facilitate the process of improving the water quality of Adayar and they stand as a primary barrier for the settlement beyond during the high flow of the river (3 SCS [2–4]).
- Constructed wetland in the lowest point on the site such that it will naturally attract the water flow and aid in indigenous method of treating Adayar [5].
- Riparian islands to deviate water from the river into the constructed wetlands.
- Biodiversity park that shall stimulate the responsiveness and the interest among citizens for promoting and protecting biodiversity.
- Micro-farming units which will engage the local residents and revive the farming activities that once occurred along the banks of Adayar near Alandur

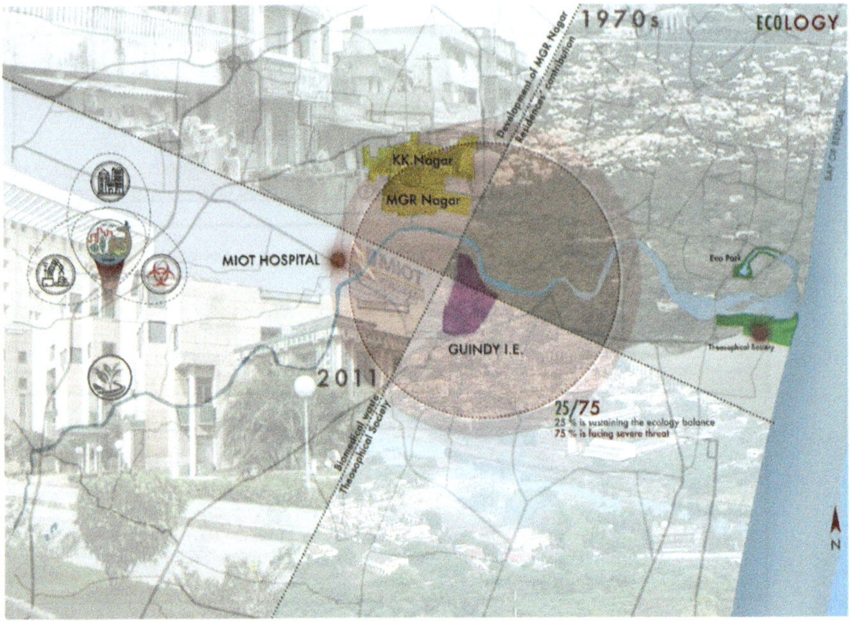

Fig. 16 Graphics illustrating the various factors affecting the ecology of the river

road. This shall be a community-based activity where the micro-farming units shall be leased to the residents on a lot basis by the government for a particular period of time and changed thereafter.

- A multifunctional event place near the Maraimalai Adigalar bridge on Saidapet side, which is a focal point with respect to Anna Salai and shall attract a large crowd when any events related to ecology and nature are conducted.
- A Boat jetty near the Tamil Nadu open university on the other side of the Maraimalai Adigalar bridge, reviving the tradition of boating.
- A cycle trail along the riparian edge toward the settlement side to facilitate easy commute between Ekkaduthangal and Saidapet/Anna Salai through the proposed riparian edge.
- Relocating the residents who are vulnerable to threats from river and also cause damage to the river bed and the water to the Guindy side, where there is availability of land (also considering the changing land use of Guindy estate and the chances that are available for the people in terms of public transport—the primary reason that attracts them to settle near the banks of the river as slums).

The proposed functions have different use currently which is either dying or not put to proper use and are stated below.

Within the critical contour where constructed wetlands and riparian edges are proposed toward the north of the river, exists low-income and middle-income housing boards constructed by the government that lacks proper infrastructure in

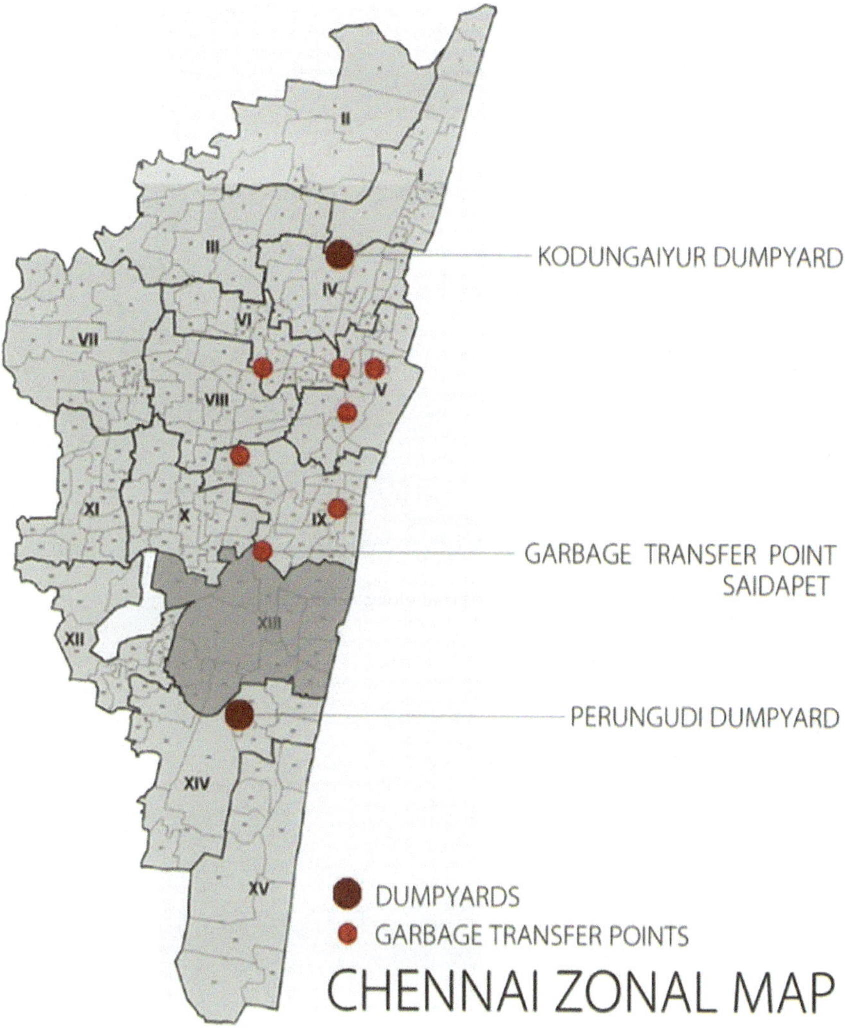

Fig. 17 Map showing the dump yards and garbage exchange points in the city

some cases and in the south of the river is vacant lands and inactive industrial buildings.

The site of biodiversity park is currently in use by slums, Survey of India office and inactive small-scale industries and warehouses.

The site of multifunctional event space is occupied by Gothamedu slum. The site of micro-farming units is occupied by cherrithottam slum:

- Stage 1—Delineating the critical contour from chosen extent (Fig. 19).
- Stage 2—Delineating resettlement zone (Fig. 20).

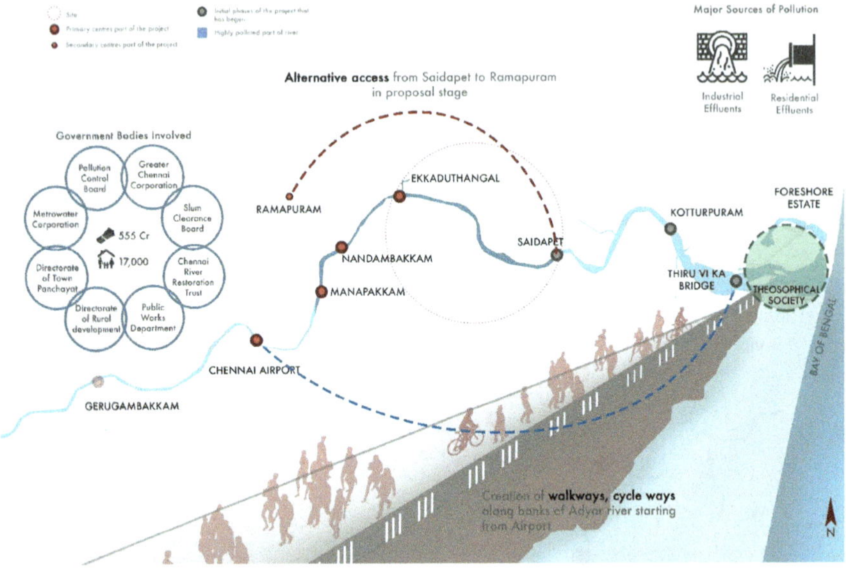

Fig. 18 Illustration of the government proposal along Adayar River

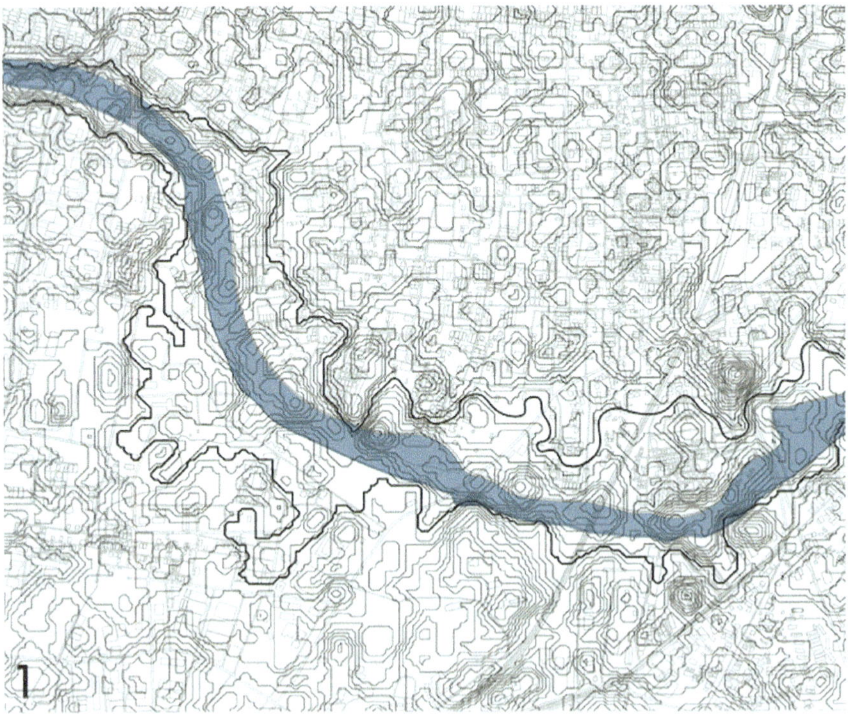

Fig. 19 Stage 1

Fig. 20 Stage 2

- Stage 3—Riparian zone along edges and within critical contour (Fig. 21).
- Stage 4—locating constructed wetlands along curvature of river to function as a parallel system (Fig. 22).
- Stage 5—Riparian islands as redirecting elements (Fig. 23).
- Stage 6—Urban farming (Fig. 24).
- Stage 7—Biodiversity park (Fig. 25).
- Stage 8—Event space and boat jetty (Fig. 26).

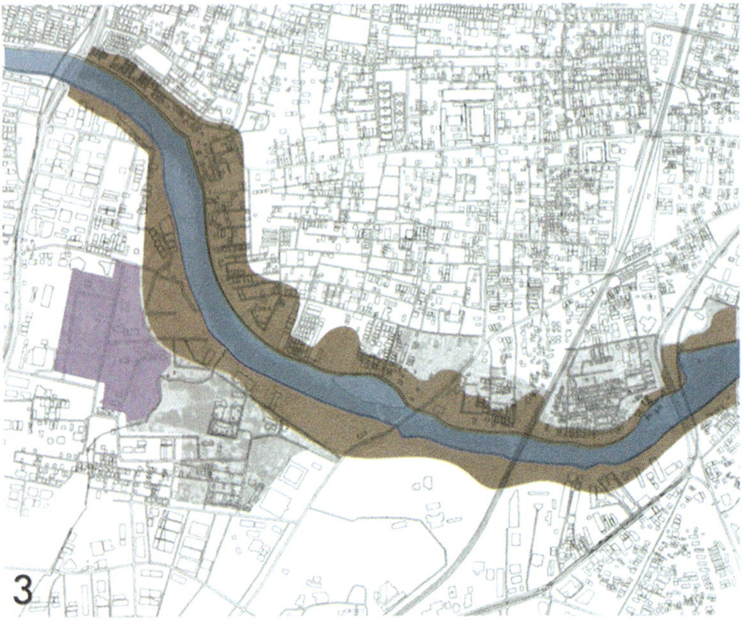

Fig. 21 Stage 3

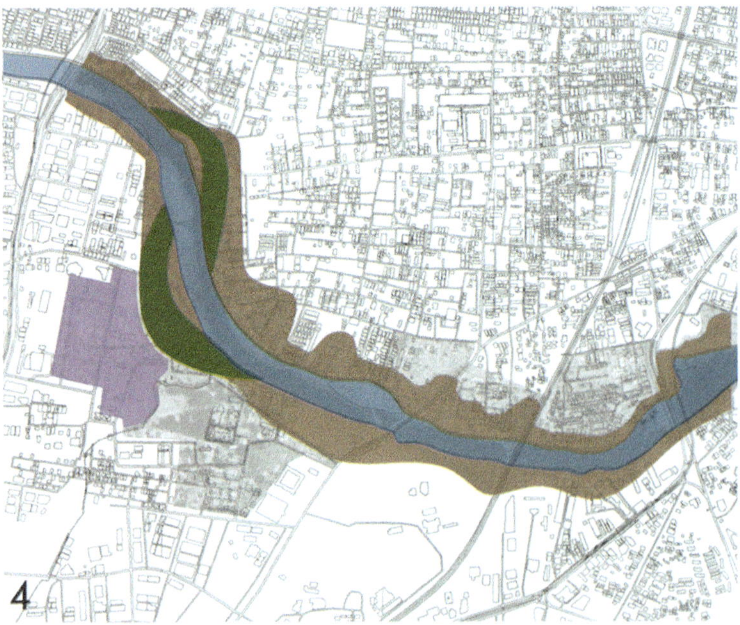

Fig. 22 Stage 4

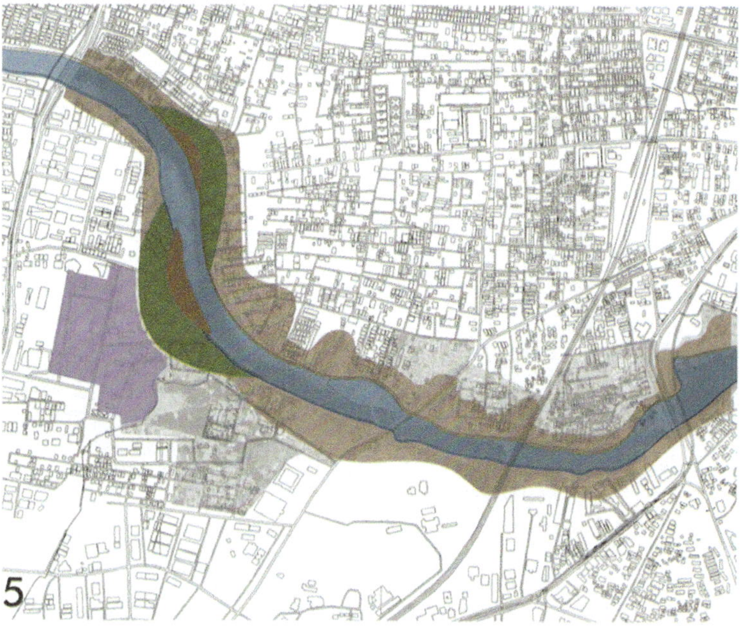

Fig. 23 Stage 5

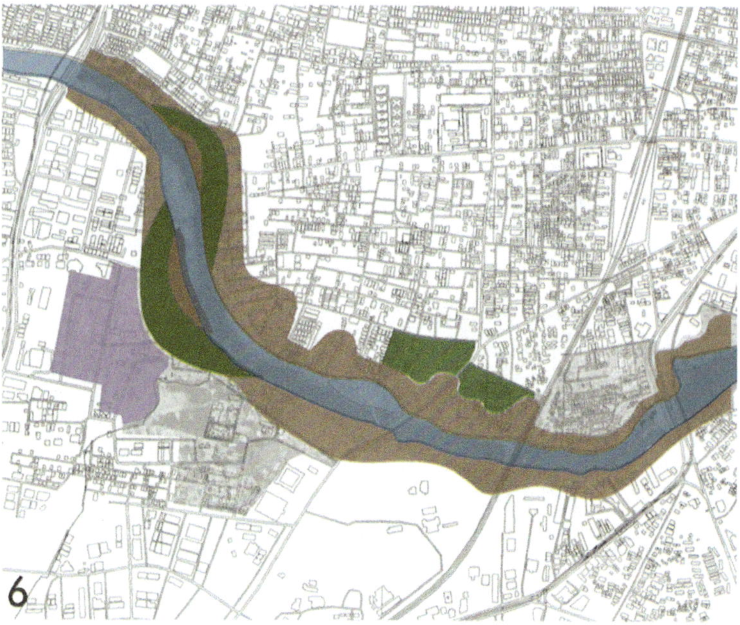

Fig. 24 Stage6

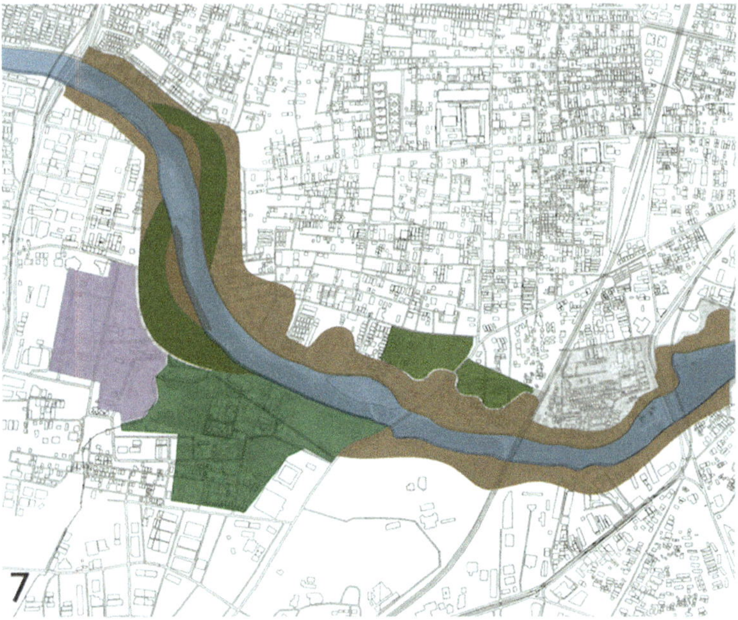

Fig. 25 Stage7

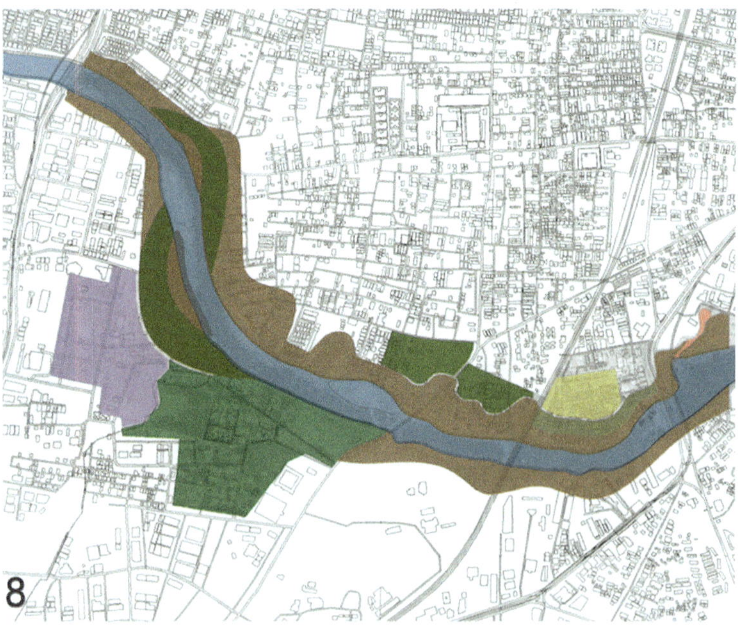

Fig. 26 Stage 8

6 Conclusion

A natural body or resource when left undisturbed shall perform its duty and shall aid to human's need and development. Now that the disturbance has occurred, we shall work toward fixing this issue and project for a sustainable future development (Figs. 27 and 28).

The recommended interventions (Fig. 21) have the capability to revive Adayar and act in response to the fast pace of climate change that is the most prevailing condition in the world. The proposal aims to be a prototype model, one which when altered according to the context of its applicability, can influence the water body and revive it and also provide a public platform-an interface of people and water, which shall have a positive impact both on the former and latter. It shall also cater to the betterment of the flood plain management and reduce the adverse effects of flood and water related disasters and react to the water crisis, hence bracing Adayar.

Fig. 27 Stage 9

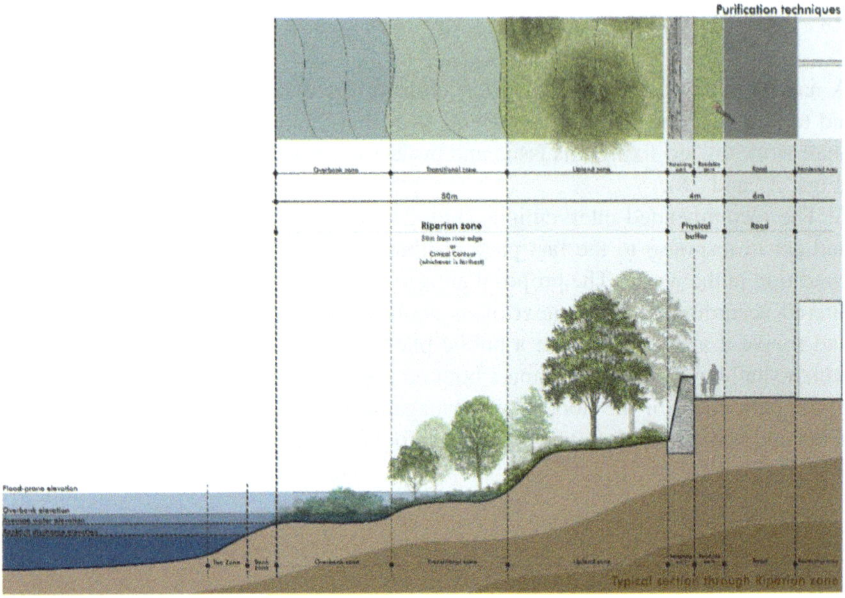

Fig. 28 Conceptual section

Acknowledgements The authors wish to acknowledge the support rendered by the professors Dr. P. Meenakumari (Asso. professor, SAP, Anna University, Chennai), Ar. Samyuktha (Assis. professor, SAP, Anna University, Chennai), Ar. Amita Gupta and Ar. Udayarajan. The authors also would like to acknowledge S. V. Divyanand, V. Laxmi, M. Surya, V. Visvesh, and G. S. Mohan for their contribution toward data collection, consolidation, and analysis work.

References

1. Vaithyanathan L, Solomon S, Priya SK (2016) A Short term study of Dissolved Oxygen behavior in Adyar River, Chennai. IJLRET 2(9), pp 43–49
2. SCS (Soil Conservation Service) (1992) Engineering Field Handbook, Chapter 13: Wetland restoration, enhancement, or creation. 210-EFH, 1/92
3. Castelle A, Johnson A (2000) Riparian Vegetation Effectiveness: Technical Bulletin No. 799. Research Triangle Park, NC, National Council for Air and Stream Improvement, Inc. 36 pp
4. Bongard P (2009) Riparian Forest Buffers for Trout Habitat Improvement: A Review. University of Minnesota Extension
5. Watson JT, Hobson JA (1989) Hydraulic design considerations and control structures for constructed wetlands for wastewater treatment. In: Hammer DA (ed) Constructed Wetlands for Wastewater Treatment. Lewis Publishers, Chelsea, MI, pp 379–391

Intelligent and Responsive Architecture

Place Identity Along Highways: Location Choice of Elements Using Distance and Isovist Measures

G. Ophylia Vinodhini and A. Meenatchi Sundaram

Abstract The aspects of physical environment help to link the dynamic understanding of people–place connect. Visibility to identity elements is a consequential aspect, notably across travel routes. It is significant to understand the degree of visibility from a vantage point along the travel routes, to define, characterize, and maximize the sense of orientation for drivers and travellers. The necessity arises to refer to the questions of a traveller or driver on the distance, spatial coverage, and enclosure of the visible spatial territories. The visibility of the volume of space together with a specification of the location of that point from the given point in space is an isovist. The isovist measures are appropriate tools to calculate, represent, and thereby develop analysis to formulate urban design strategies based on the visual aspects. This paper aims at identifying the relationship between the location choice and the visual access to the identity elements along a highway. Distance-based location choices are studied to generate relationships between distance, speed, and travel time in three criteria. Evaluation of Visibility with the help of a simulation tool [1] in two forms of urban nodes with significant natural, cultural, and social elements, resulting in 32 isovist maps are generated with 16 maps for each node. The degree of visibility concerning the openness from highway and placement of elements are extractions from the analysis. The results show that the visibility of elements and the location characteristics contribute to the awareness of place identity along travel routes.

Keywords Urban nodes · Visibility index · Isovist · Highways · Location choice · Place identity

G. Ophylia Vinodhini (✉) · A. Meenatchi Sundaram
Department of Architecture, NITT—National Institute of Technology, Tiruchirapalli, Tamil Nadu 620015, India
e-mail: ophyliagodwin@gmail.com

G. Ophylia Vinodhini
C.A.R.E. School of Architecture, Tiruchirapalli, Tamil Nadu 620009, India

K. Thirumaran et al. (eds.), *Sustainable Urban Architecture*, Lecture Notes in Civil Engineering 114,
https://doi.org/10.1007/978-981-15-9585-1_10

1 Importance of Identity Elements Along Travel Routes

Any travel experience has a positive psychological motivation to understand the people, place, and environment which a traveller passes by [2, 3]. Human beings have the curiosity to access their places of interest. The relationship between people, travel destinations, and the identity elements alongside routes inextricably intertwines to manifest the values and identity in travelling [4]. Travellers of today are witness to the disappearance of identity elements sought along roadscapes to experience the heritage, picturesque environments, distinctive landscapes, and cultural societies [5]. There is a present need to address place identity elements as an essential area of research to enrich the roadscape environment with awareness and orientation for the travellers. Any place that reflects the identity of a person or place and rests in the memory of people has place identity [6–8].

Urban identity is the physical and perpetual features of the urban space that are the combination of culture, social structure, and the needs and functions of a built environment. The physical elements and activities have a strong influence on the users' identification of the attributes of a particular place [9]. These elements become place markers in people's mental map about a particular place [10]. They also convey the historical messages, the evolution through times, and the contemporariness of present-day context. Irrespective of the permanency or temporariness of these elements, they build a strong sense of society, individuality, uniqueness, and distinctiveness through the cultural expression across timelines. The absence of these elements usually occurs due to uniform concepts in homogeneous planning which would gradually lessen the local identity [11]. The weakening of identity can result in the loss of meaning and emotional attachment associated with the context. The lack of meaning decreases the understanding of the place to its real sense of place and identity [12].

1.1 Place Identity Elements Along Arterial City Roads and Highways

The imageable city roads with legible identity help to build a strong relationship between human behaviour and their surroundings. They are thereby helping the travellers to recognize and recall a place. For ages, people's conceptions and value systems have evolved and been transformed, especially in the traveller's experiences along the country's highways [13, 14]. The society living in the current decade and past two decades is witness to sudden transformations in conveyance with high-speed roads allowing people to reach their destination at unprecedented speed in the comfort of their conveyances. The new proposals and the widening of roads often undergo acquisition of lands. Developers and entrepreneurs in developing countries visually perceive this trend as giving them additional access routes and visibility to establish their businesses along these roads. Geometric design with safety was the established criteria for many decades [15–17]. People's emotions, memories, and nostalgia are often the unseen parts, which are unnoticed in these developments. Many places change, often losing the important place markers and elements. Place association becomes insignificant as these changes happen, and the mundane attitude of travel routes springs up to disinterest the traveller.

1.2 Travel Route as an Inspiration

Earlier, the travel corridors of culturally rich places were prominent for certain elements reflecting the cultural identity of the place, but the roads of today deprive of these reflections that once delighted the natures of traveller. Every structure, tree, and other elements were connectors of the identity of a place and accessible from the travel route and contribute something paramount to the experience of place identity. Understanding the pragmatic aspects of this concept is a key to addressing the difficulties in examining the history of the roadside landscape, bygone travellers' feelings of place identity, and momentary elements along the present travel routes that are eradicating the unique character of intercity areas.

The burgeoning car traffic is the crucial segment that depleted urban life out of the urban space. The growth in vehicular traffic has numbed our senses to the contextual approaches and observations around us. A road that inspires people to travel should offer places of interest with spatial elements to stop, look, learn, relax, or get entertained. Such elements can help to increase the activity engagements and the understanding of awareness and orientation along the travel routes. There are always more eyes on these roads that inspire to follow the places and elements passing by. The public spaces and transport system are well connected, yet have their requirements spatially and functionally as two sides of a coin [18]. Building more roads is always an incitement to buy more cars, which eventually increases the desire to travel and drive to faraway destinations. These long drives call for an interest in the exploration of the surrounding environments.

1.3 Locating Identity Elements

Sensory development closely ties to evolutionary history and simply classifies into the distance senses [19]. Sight is the most highly developed of our senses. Locating the elements as shown in Fig. 1, with clear visibility, is a valid logical positioning to attract the traveller. Permanent monuments make an impact, but temporary sculpture exhibits and specific thematic installations can have a more considerable influence on the observer. The concentrated view can contribute to a sense of identity. The uniform concepts and homogeneous planning may lessen the local identity.

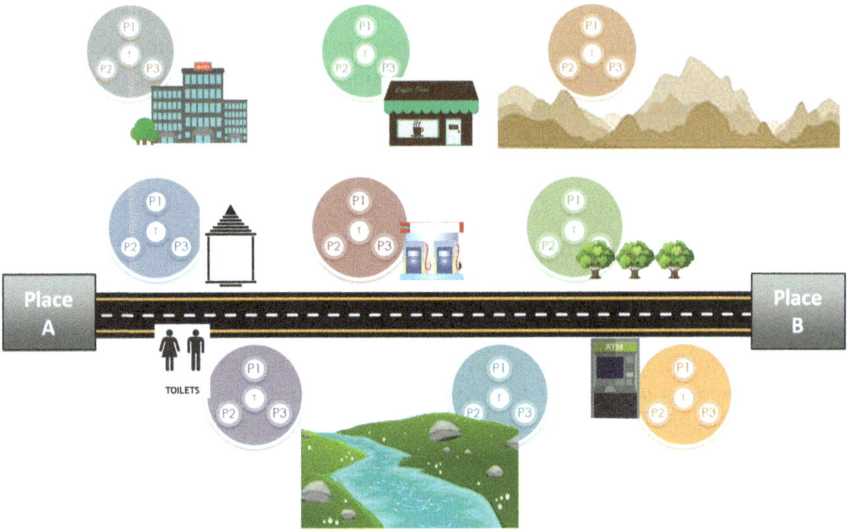

Fig. 1 Identity elements along roadsides

2 Visibility Access: Speed of Travel and Cone of Vision

The predominant perceptions of urban fabrics occur through visual observation. The visual fields are essential to human perceptions and apply to animal instincts. Specific indicators improve the awareness of people by providing access to mentally localize and record the presence of remarkable elements to orient oneself in any particular place. According to Lynch, every human being continuously involves in the efforts to organize ourselves in the situations and surroundings where we exist, most often concerning a built structure or notable elements around us that we identify in particular or with connections [20]. Visibility access is the condition where the user can perceive a space but not be able to traverse to that space. The signage in an urban space indicates the presence and direction of a space or group of spaces. These reduce the conceptual distance and link one space to another. The frame of visibility focuses on how people see the place concerning the identity of a particular place [21]. Figure 2a depicts the cone of vision of a driver's perspective with distances of vision and Fig. 2b the travellers' cone of vision as they see from the sides.

2.1 Isovist and Isovist Measures

A review of visibility analysis techniques applied to architecture and urban environment is documented comprehensively by various researchers [22, 23]. Syntax

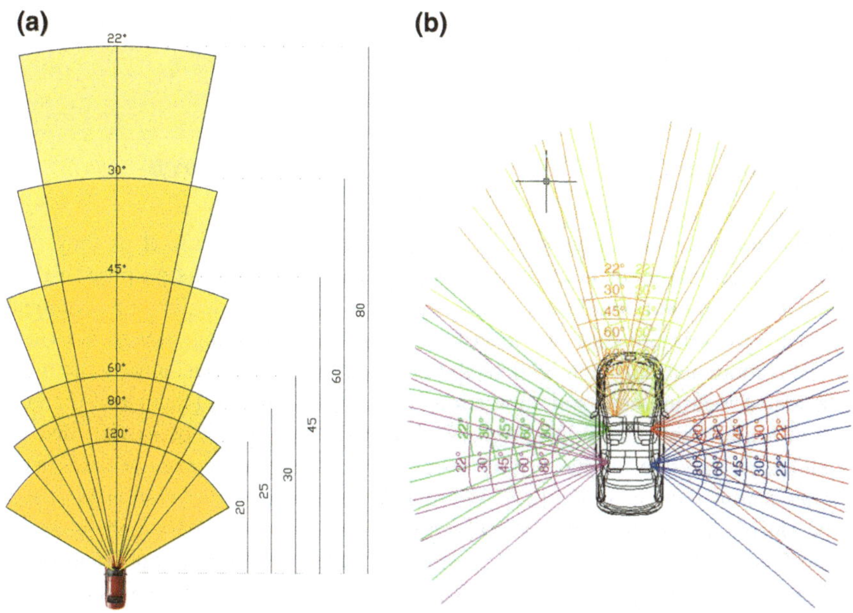

Fig. 2 Cone of vision—visibility angles: (**a**) driver's vision and (**b**) travellers' vision. Measurement and simulation methods for visibility analysis

2D software calculates using grid analysis and the two paths in the fabric to analyse different urban forms. Previous researches show that different visual qualities exist in different urban forms concerning their paths [24]. The walls, buildings, and trees in an urban context are to create an experience of people who move through and populate the cityscape. The visibility field analysis is traditionally the viewshed analysis to analyse as per the angles of vision. The 2D bounded polygon defines the open space concept. The open spaces are empty or void space between the buildings. The key questions in viewshed analysis include the distance, area of visibility, and space enclosure of visibility field, which are important parameters in urban design studies [25].

The axial line of view that covers a convex space and the cone of vision that forms the isovist field are some of the ways to understand the visibility of any space at any given point. An isovist field is profoundly affected by changes in the direction of building arrangement or partial transformation of shape. In isovist field analysis, the space to be analysed is divided into the desired number of grids, forming isovist figures from the central point of each grid and analysing geometric characteristics to identify overall visibility characteristics [22]. Hillier focused on structural patterns through visual connectivity of the space [26]. The axial and convex maps applied in space syntax divide a space into formal units and convert them to a mathematical graph for path search. In visibility graph analysis, the axial lines, convex spaces, and isovist fields are into grid divisions. The visual information in a high-density urban environment with full of high-rise buildings is affected by area, shape, and direction of building façade.

2.2 Application of Isovist Approach in This Research

Identity induced by noticeable elements along highways is crucial as they contribute to orientation, awareness, identity, and sense of place. The purpose of these elements is beyond wayfinding, also being vital to the legible environment. Physical environs are one of the critical factors in impacting human activities. Isovist is more appropriate to understand the visibility boundaries, among the many methods of measurement and simulation [22, 24, 27]. Gibson introduced the concept of 'optic flow' in the landscape [28]. The shape and size of isovist filed are essential, as they are unique to the observer's position. A change in isovist field implies a transition in space [27]. Wayfinding is a series of execution processes [29]. During navigation, the user interacts with the environment and gains his knowledge to find the best strategy. The awareness of orientation, memory of landmarks, ages, and genders, may all contribute to various wayfinding behaviours. The most stopping points are at intersections near the concave shape and have relatively smaller isovist areas or in transitions from big to smaller areas. In spaces with small isovist areas, the observers have limited information about the area. When an observer is navigating, the spatial behaviours can seem like a mere response to the transformation to invariant structures of the environment, based on one's ability to perceive and experiences [30].

3 Research Methods

The method to decipher the location choice of identity elements depends on the origin and destination points and the distances between them. Biological needs to stop/pause and the travel duration are major decisive factors. A structural and spatial model of a typical road with the roadside lane segments is the typical scenario of the study and is into four sub-corridors. The understanding of equations in physics using speed, time, and distance with calculations of suggestive pause points helps to derive the equations for the location of pause points. The visibility of elements can be predicted using calculations and simulations. Two nodes of significance arrived from previous academic studies are chosen. Node 01 has a 1200-year-old temple of archaeological importance, situated on a picturesque rock at less than 60 m from the highway. Node 02 has a canal, which is an irrigation source constructed around 1000 AD in the later Chola period. Both the nodes are of historical, cultural, natural, and functional importance. UCL Depthmap is suitable software to understand the visibility using visibility graph analysis (VGA) and agent analysis (A.A.). A graph generated with the isovist areas at 16 conditions for each node is the outcome of the simulation in depthmap. Besides, are eight maps on either side of the road, with four mapping angles of isovist—90°, 120°, 180°, and 360°. Figure 3 shows the methodology. The discussions of the results follow in the further sections.

4 Results and Discussion—Locating Identity Elements

4.1 Distance-Based Locations

Figure 4 shows the lanes and the positions of typical roadside elements, where W1 is right of way, W2 is the adjacent lane to the right of way, W3 is lane next to W2, and W4 is lane next to W3.

$$W_{cor} = W_1 + 2W_2 + W_3 + W_4 \tag{1}$$

Equation 1 shows the width of the corridor as a sum of twice the width of sub-corridors abutting the road added to the width of the right of way (Table 1).

Equations 2 and 3 are multiplied with the fractions of half and two-third to get the pause point. Equations 4 and 5 generate with the number of interesting stops required to the traveller to pause to observe and appreciate an identity element along the travel route.

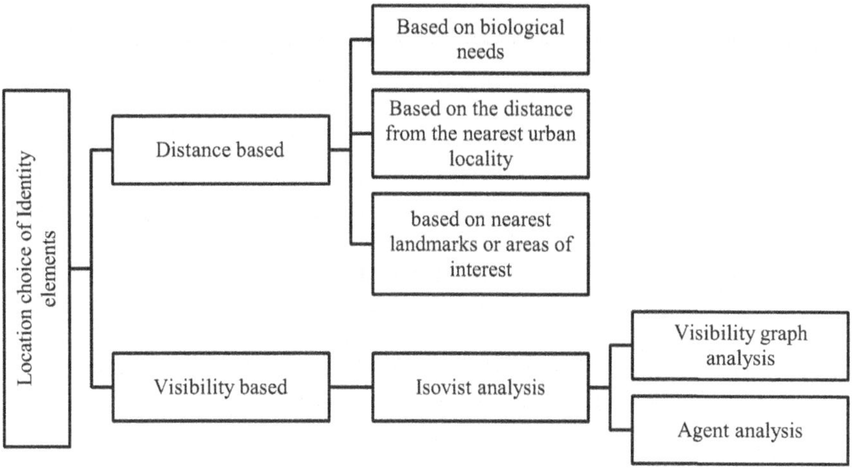

Fig. 3 Methodology framework

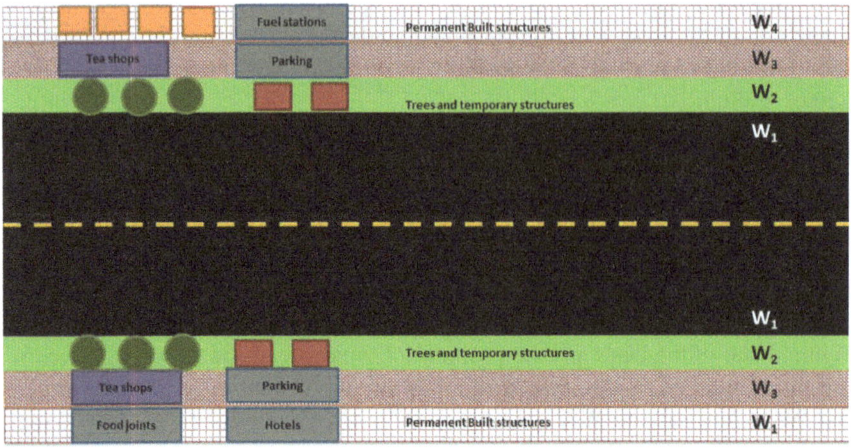

Fig. 4 Width and lanes-based locations of identity elements

Table 1 Equations for stopping distances along travel routes

Criteria	Equations	Explanations
Biological rest limit—½ of travel	$DPb_1 = \frac{1}{2} T \cdot S_{avg}$ (2)	DPb_1 and DPb_2 = Pause point T = Travel time duration S_{avg} = Average speed of travel
Biological rest limit—2/3 of travel	$DPb_2 = \frac{2}{3} T \cdot S_{avg}$ (3)	
The distance from the urban area situated nearby	$DP_u = \frac{L_T}{N_S + 1}$ (4)	DPU and DPC are the distance between stopping or pausing points LT = Total length from the origin to the nearest urban area. N.S. and NCE = Number of stops required
The distance from the landscape element located nearby	$DP_c = \frac{L_T}{N_{ce} + 1}$ (5)	

Source Author's derivation using basic equations of time, distance, and speed

4.2 *Visibility Analysis—Isovist Tool*

The visibility graph analysis (VGA) clearly shows the areas of visibility in yellow, and blue is the region where moderate visibility occurs. The higher red hue denotes the larger isovist area and thereby broader visibility. Figures 5a and 6a shows the visibility graphs generated for the Nodes 01 and 02. Point isovist tool along the road is to understand the visibility at a point along with the travel. VGA visualizations have increased colour depth and reference scales. The agent analysis shows the movement pattern of virtual people in the environment [31]. This analysis requires the visibility graph and uses occlusions and length of the line of sight to generate the map, as shown in Figs. 5b and 6b. Further, Figs. 7 and 8 show the 32 images generated using the isovist tool for Nodes 01 and 02.

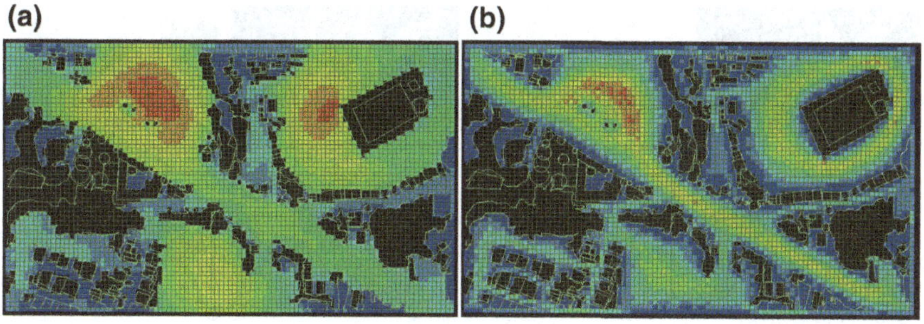

Fig. 5 Node 01—(**a**) visibility graph analysis and (**b**) agent analysis

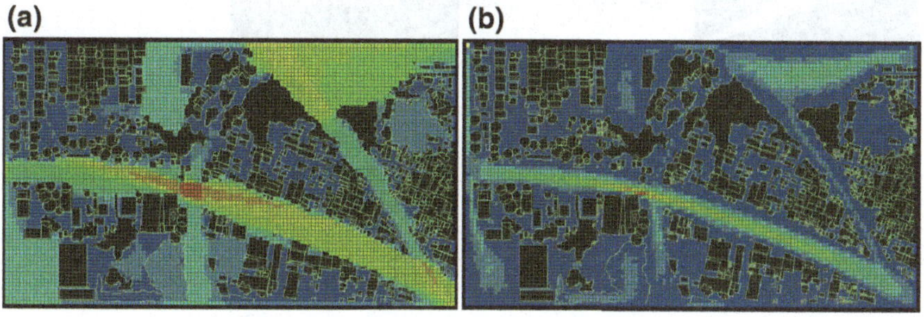

Fig. 6 Node 02—(**a**) visibility graph analysis and (**b**) agent analysis

The study shows the applicability of isovist tool in analysing the location choices of identity elements along highways. Two nodes with two different characteristics to evaluate visibility using depthmap prove the degree of visibility index is relatively high due to lesser obstructions in Node 02 but low due to eye-level obstructions in Node 01. The studies and analysis show that those area indices of isovist have a more substantial influence on the visibility of any element. The water body in Node 02 has lesser visibility than the rocky temple in Node 01. The characteristics such as horizontality and the edges of water body confine than the large rock temple that is more visible from the highway.

Figure 9 illustrates the degree of visibility in terms of isovist areas with the graph drawn from the simulations generated with the isovist areas from visibility graph analysis (VGA) and agent analysis (A.A.) as simulated from the software Depthmap X0.50 [26, 32, 33].

The observations are that the VGA analysis shows more visibility than the agent analysis. As an outcome, point P3 (120°) of Node 02 shows the highest visibility and P4 (120°) of Node 02 shows the least visibility. In the agent analysis results, P4 (120°) shows the highest visibility and P2 (90°) shows the least visibility. However, the vision angle 120° has more coverage of isovist areas.

Fig. 7 Node 01—isovist mapping generated at 90°, 120°, 180°, and 360°

Fig. 8 Node 02—isovist mapping generated at 90°, 120°, 180°, and 360°

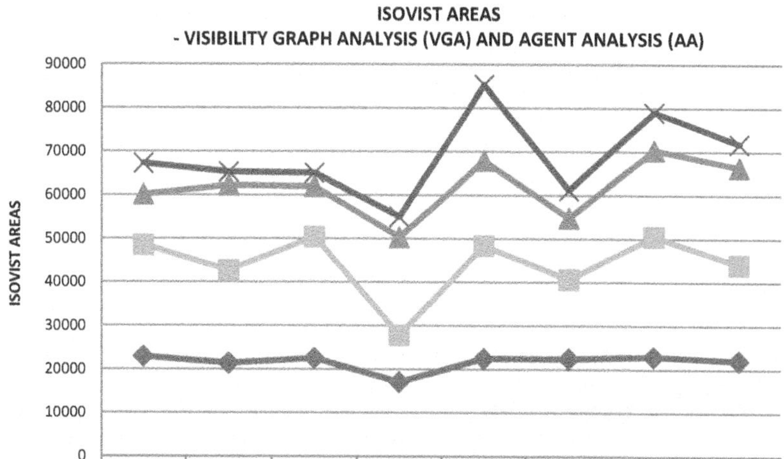

	P1(90)	P2(90) opp	P3 (120)	P4 (120) opp	P5 (180)	P6 (180) opp	P7 (360)	P8 (360) opp
Node 02 AA	7105.572	3022.837	3105.987	4719.957	17451.676	6381.7822	8721.3193	5401.4272
Node 01 AA	11600.991	19519.037	11484.636	22479.678	19492.412	14200.537	19785.309	22366.49
Node 02 VGA	25611.25	21280.627	27904.123	10698.031	25930.047	18139.438	27566.682	21949.98
Node 01 VGA	22872.982	21385.496	22474.666	17084.996	22431.527	22422.998	22919.008	22016.797

POINTS OF ISOVIST OBSERVATIONS

Fig. 9 Graph depicting isovist areas of Nodes 01 and 02

5 Conclusion

Isovist is a systematic procedure to identify visibility along highways. The critical finding of the research is the isovist angle 180° provides larger isovist areas, thereby greater visibility, whereas the lesser visibility occurs in the isovist angle 120°. Similarly, more number of stopping points occurs in scenic and cultural highways with more elements of identity. The least elements occur in the mundane highways of arid areas. It becomes essential to integrate the visibility and character of a highway to locate the appropriate elements at appropriate pause point locations considering the distance, speed, and duration of travel.

The model suggests visibility extents, sufficient to plan and locate identity elements. The analysis of physical and biological features helps to estimate the possible visual space. It also shows the limitations and obstructions, which can help improve the skyline, access to the elements, and enhance the focus of visibility. Though the method technically represents the visibility, specific additional methods would be necessary to understand the cultural experiences and personal values of travellers as well as people of the place. The transparency and flexibility of the depthmap help in achieving the results faster for a quick analysis. The measurable data is an initial step to establish the link between visibility and location choices. Finally, the study deciphers the importance of visual access and the appropriate location of identity elements. The softwares to develop three dimensional isovists are still in development and can help to improve the understanding of visibility enclosure in the third dimension as the future of this type of researches. The layering of visibility with characters of identity elements is another direction for this research. Isovist analysis could become an inherent part of site analysis for design projects involving built environment and landscapes.

References

1. DepthmapX development team (2017) *DepthmapX (Version 0.50)* [Computer software]. Retrieved from https://github.com/SpaceGroupUCL/depthmapX/
2. George BP, George BP (2004) Past visits and the intention to revisit a destination: Place attachments as the mediator and novelty seeking as the moderator. J Tour Stud 15(2):51
3. Kolodziej A (2017) Landscape routes as an infrastructural core of cultural landscapes; Their distinctive role for the character of region. In IOP Conf Ser: Mater Sci Eng 245(4):042053. https://doi.org/10.1088/1757-899X/245/4/042053
4. Anggraini LM (2015) Place Attachment, Place identity and Tourism in Jimbaran and Kuta, Bali (Doctoral dissertation, University of western Sydney). http://handle.uws.edu.au:8081/1959.7/uws:32139
5. Gieseking JJ, Mangold W, Katz C, Low S, Saegert S (Eds) (2014) The people, place, and space reader. Routledge. https://doi.org/10.4324/9781315816852
6. Najafi M, Shariff MKBM (2011) The concept of place and sense of place in architectural studies. Int J Hum Soc Sci 6(3):187–193
7. Lewicka M (2011) On the varieties of people's relationships with places: Hummon's typology revisited. Environ Behav 43(5):676–709. https://doi.org/10.1177/0013916510364917
8. Killguss BM (2008). Identity and the need to belong: Understanding identity formation and place in the lives of global nomads. Illn, Cris & Loss 16(2):137–151. https://doi.org/10.2190/IL.16.2.d
9. Blumentrath C, Tveit MS (2014) Visual characteristics of roads: A literature review of people's perception and Norwegian design practice. Transp Res Part A: Policy Pract 59:58–71. https://doi.org/10.1016/j.tra.2013.10.024
10. Fredericks B (2013) Trees along our travelling tracks. About Place J 11(1):1–5
11. Charlton SG, Mackie HW, Baas PK, Hay K, Menezes M, Dixon C (2010) Using endemic road features to create self-explaining roads and reduce vehicle speeds. Accid Anal Prev 42 (6):1989–1998. https://doi.org/10.1016/j.aap.2010.06.006
12. Hull RB IV, Lam M, Vigo G (1994) Place identity: symbols of self in the urban fabric. Landsc Urban Plan 28(2–3):109–120
13. Clements J, Swaffield SR, Wilson J (2010) Landscape and Associated Environment Values in the roadside corridor: A selected literature review. Lincoln University. LEaP. URL: https://hdl.handle.net/10182/580
14. Silberberg S, Lorah K, Disbrow R, Muessig A (2013) Places in the making: How placemaking builds places and communities. Massachusetts Institute of Technology, Department of Urban Studies and Planning, Cambridge, MA, USA
15. Boarnet MG (2014) National transportation planning: Lessons from the U.S. Interstate Highways. Trans Policy 31:73–82. https://doi.org/10.1016/j.tranpol.2013.11.003
16. Czynska K, Rubinowicz P (2015) Visual protection surface method: Cityscape values in context of tall buildings. In: 10th international space syntax symposium (p 142)

17. Mackie HW, Charlton SG, Bass PH, Villasenor PC (2013) Road user behavior changes following a self-explaining roads intervention. Accid Anal Prev 50:742–750. https://doi.org/10.1016/j.aap.2012.06.026
18. Pilat-Borcuch M (2014) Intangible Qualities of Good Urban Design. Geography 49(1)
19. Copeland WE (2013) Addressing Local Development and Local identity: Rethinking the Chapman Highway Corridor in South Knoxville. https://trace.tennessee.edu/utk-gradthes/2405/
20. Lynch K (1965) The view from the road. Ann Assoc Am Geogr 55(4):629
21. Dalton NS, Dalton RC (2010) Solutions for visibility-accessibility and signage problems via layered-graphs. J Space Syntax 1(1):164–176
22. Chang DK, Park JH (2011) Explanation of isovist fields to model 3D visibility with building façade. Arch Res 13(3):19–29. https://doi.org/10.5659/AIKAR.2011.13.3.19
23. Franz G, Wiener JM (2008) From space syntax to space semantics: a behaviorally and perceptually oriented methodology for the efficient description of the geometry and topology of environments. Environ Plan B: Plan Des 35(4):574–592. https://doi.org/10.1068/b33050
24. Alalhesabi M, Hosseini, SB, Nassabi F (2012) Housing visual quality in urban pattern Application of isovist method in old fabric of Bushehr city. Iran Univ Sci & Technol 22 (1):60–64
25. Wolosyn P, Leduc T (2011) A landscape potential characterization: spatial template of pedestrian ambient fields within the urban fabric
26. VaroudisT, Law S, Karimi K, Hillier B, Penn A (2013) Space syntax angular betweenness centrality revisited. In: Ninth International Space Syntax Symposium Seoul: Sejong University (No. 057, pp 1–16)
27. Chen CH, Lin TJ, Chen CY (2016) From Isovist to Spatial Perception: Wayfinding in Historic Quarter. Environ-Behav Proc J 1(3):300–310. https://dx.doi.org/10.21834/e-bpj.vli3.374
28. Weller G, Schlag B, Friedel T, Rammin C (2008) Behaviorally relevant road categorization: A step towards self-explaining rural roads. Accid Anal Prev 40(4):1581–1588. https://doi.org/10.1016/j.aap2008.04.009
29. Troffa R (2010) Visibility and way finding: a V.R. study on emergency strategiesEnviron Model: Using Space Syntax Spat Cogn Res, 7
30. Won S, Kim S (2017) Mobility is in the eye of the beholder: A comparison of travel patterns and urban spatial use between migrants and the original residents of Danang, Vietnam. Cities 67:63-73. https://doi.org/10.1016/j.cities.2017.04.01
31. Varoudis T, Penn A (2015) Visibility, accessibility and beyond: next generation visibility graph analysis. In: SSS 2015—10th International Space Syntax Symposium
32. J.Pinelo and A.Turner, 'Introduction to Depthmap', 2010
33. Gil J, Varoudis T, Karimi K, Penn A (2015) The space syntax toolkit: Integrating depthmapX and exploratory spatial analysis workflows in QGIS Proc 10th Int Sp Syntax Symp, p 148

Built Environment

Envelope Performance Analysis of Office Buildings in Warm and Humid Climate: From Case Studies of Multi-storied Office Buildings in Chennai

Chandrasekaran Chockalingam

Abstract In India, the average of all commercial buildings' electricity consumption is at 6.6% and the growth of office spaces in urban area is increasing at the rate of 8–10% annually since 2005 [1]. India is still in the nascent stage of energy conservation and energy-saving potential is 20–40% for commercial buildings [2]. Hence, the Bureau of Energy Efficiency (BEE) has formulated the Energy Conservation Building Code (ECBC 2017) which primarily stipulates codes for the design of energy-efficient buildings. The purpose of this research paper is to explore the energy performance of six office building samples from each typology of the predominant envelope design in Chennai since 1995 and analyze the sample buildings for three major envelope performances—solar insolation, daylighting and space-cooling load. The first part is performance ranking from site measurement data and then checks for ECBC 17 compliance for all the samples. As the age of HVAC equipment and other space-cooling components varies in all the samples, the results were inconsistent to compare; hence, a simulation method was adapted by creating building information model in Revit using same envelope building materials. Comparing all the case studies, the results for the case buildings with energy-efficient façade design performed better than the rest. Based on the analysis and results, this paper lists the envelope design factor for of multi-storied office buildings in warm and humid climate of Chennai.

Keywords Envelope performances factor · Energy simulation model · Energy efficient envelope · ECBC 2017

1 Introduction

Chennai is one of the largest markets for office space in India [3] with significantly rapid growth in the past two decades. In addition, it is crucial to focus on energy-efficient building envelope design in Chennai, where space-cooling load is major energy consumption for this climate.

C. Chockalingam (✉)
Faculty of Architecture, Dr. MGR Educational and Research Institute, Chennai, India
e-mail: chandru.motif@gmail.com

K. Thirumaran et al. (eds.), *Sustainable Urban Architecture*,
Lecture Notes in Civil Engineering 114,
https://doi.org/10.1007/978-981-15-9585-1_11

Energy-efficient building envelope is core objective of façade design of the twenty-first century as they influence more than 50% of the cooling load for fully glazed buildings with window-to-wall ratio (WWR) more than 70% [4]. National Action Plan on Climate Change (NAPCC) is a government policy document prepared by Prime Ministers' council on climate change that ambitious goal set to achieve energy-saving measures. It also mandated energy saving of more than 20% in building sector by 2030 [5]. Integrating thermal, daylight and energy performance for a building envelope with optimization into the design process has always been a challenge for architects. With increased sophistication on the digital tools in assessing building performance, a great potential exists to optimize the performance of contemporary building. It is possible to build super-efficient buildings at little or no added cost using current technology, but to achieve these targets it is imperative that there is a systemic change in the approach of design and development of commercial buildings. In Chennai, the typology of office building envelope design up till 1995 with recessed window is shown in Fig. 1. However, in 2000 the need for more information technology (IT) professional and service demanded bigger IT office spaces like the Tidal Park (2001), Fig. 1. In 2005 master plan, the Chennai Metropolitan Development Authority (CMDA) revised the multi-storied building bylaws for floor space index (FSI) more than 3.0. The archaic and decaying industrial estates with large plots in Ambattur, Guindy and other area saw this as an opportunity and transformed to rented IT office parks as in Fig. 1. The office market has experienced a remarkable growth from 10 million sft in 2005 to over 40 million sft in 2011 [6]. Office built for the IT industry constitutes 86% of the office space stack in Chennai, and it is one of the India's leading energy-efficient cities with more certified green buildings [7].

1.1 ECBC 2017 Overview

India's modern office uses over twice the amount of energy of the international counterparts [8]. Energy use intensity (EUI) for Indian office is a range from 200 to 400 kWh/sqmt/yrs whereas in international office buildings EUI of <150kWh/sqmt/yrs. To reduce the energy and aim for positive economic development, BEE has formulated the energy-efficient building code initiated in 2002. The latest ECBC

Fig. 1 Office buildings in Chennai: (**a**) LIC tower (1960), (**b**) Meena Kampala Tower (1999), (**c**) Tidal Park (2001), (**d**) Olympia Tech Park (2010), Guindy, (**e**) Indiabulls Tech Park (2012), Ambattur. *Source* Author

2017 prescribes a minimum standard for energy use in new and major retrofits of buildings. Tamil Nadu among the eight states has notified to apply this code soon [9].

1.2 Methodology

All sample building and daylight data are calculated manually as per ECBC manual method and the annual energy consumption history from Tamil Nadu Electricity Board or building operation and maintenances. The samples are ranked based on its envelope performances and checked for ECBC compliance as in Fig. 2.

Based on the site measurement data, a BIM was created using Autodesk Revit 2019. Simulating daylight and energy performance is done using Insight (a plug-in for Revit). These data are used to compare the results for sample buildings to obtain the envelope performance factors. These results were analyzed to establish the methodology for future integrated approach.

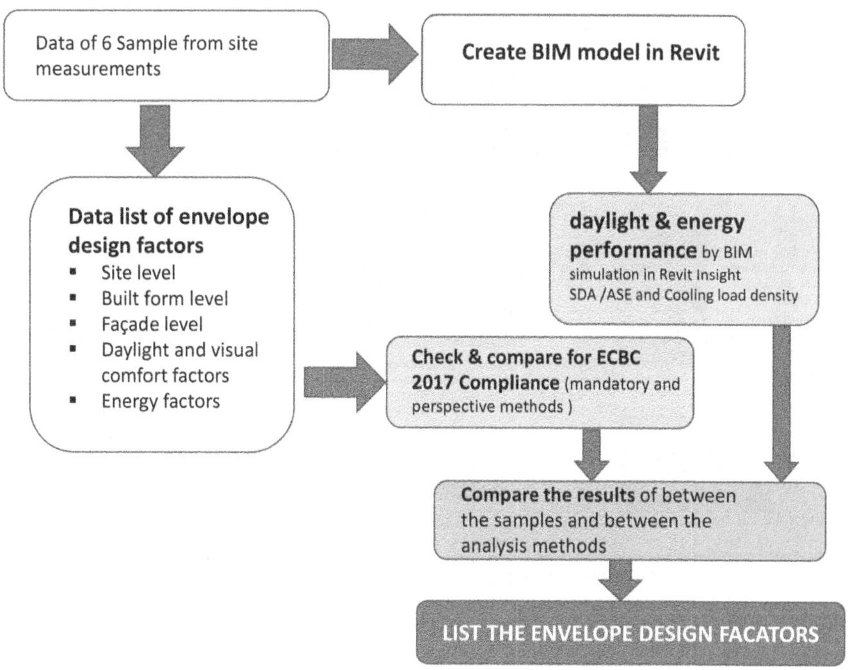

Fig. 2 Methodology flowchart

1.3 Selection Criteria for Six Sample Office Buildings

Only office use buildings, multiple tenants, electricity load more than 100KW and Annual electrical consumption data is available are the criterias for selection of case samples. Each sample has its own uniqueness and represents the typical typology of façade design of Chennai as in Fig. 3.

The physical data from site measurement in the following categories are (Fig. 4):

- Site level
- Built form level
- Envelope level
- Daylighting performance
- Energy use.

2 Envelope Performance Ranking

Rank the samples as per their performances for built form, envelop, daylighting and energy use by ranking weightage system. The ranking parameters as mentioned in Fig. 5.

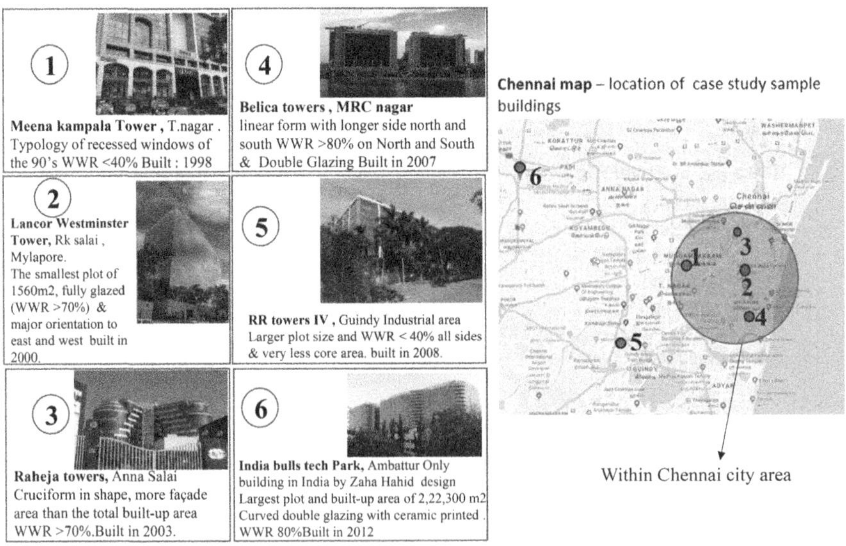

Fig. 3 List of sample buildings from samples 1 to 6 and its location map buildings in Chennai. *Source* Author

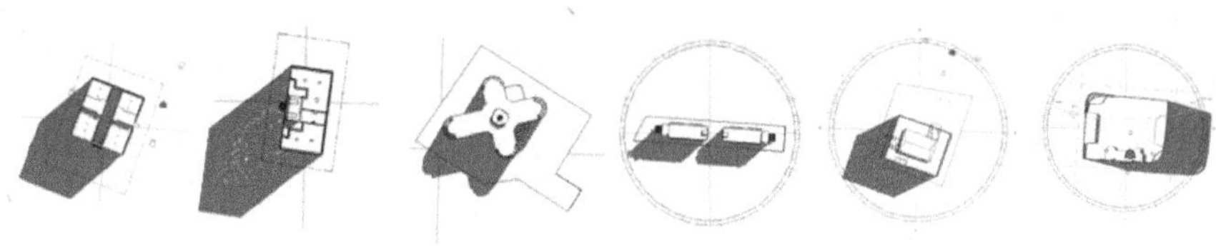

1 2 3 4 5 6

Fig. 4 Top view of shape of the built form for all samples from 1 to 6. *Source* Author

Observations

- Built form and envelope level: Sample 6 performed the best, followed by samples 4 and 5, and worst performance is sample 2.
- The recessed window sample 1 performed better than sample 3.
- Daylight and energy: Sample 3 is the best as the HVAC equipment is new, followed by samples 4 and 6. The worst performance was samples 1 and 2 for its age of HVAC equipment and facade design.

2.1 ECBC 2017 Benchmark Method

Energy-efficient code compliance has type of building use and size. The sample case studies are commercial use and all three scales (small, medium and large). The building envelope compliance for commercial building will be core and shell building method. To pass the compliance, the building should pass the mandatory provisions as mentioned below and any one of the following perspective provisions [10].

2.2 Mandatory Provisions

As in Fig. 6 and Table 1.

2.3 Prescriptive Provisions

2.3.1 Prescriptive Requirements

Prescriptive requirements are same as mandatory requirements with additional requirement of WWR <40% for commercial building in warm and humid zone.

		Site building data & performance ranking	CASE SAMPLES					
			1	2	3	4	5	6
site	1	**Site area in sqmt** - area of the site	1570	3749	13377	8789	5388	41480
	2	**Total gross area - sqmt**	5830	10692	42123	43673	24675	222965
	3	**Site orientation** : orientation to true north	W- E	S	Square	S:W	N -S	N -S
building level	1	**Shape** : as in Fig 4	rectangle	rectangle	cruciform	linear	rectangle	fillet rectangle
	2	**Form orientation** - deg Orientation of the blg to true north	0	-16	15	4	18	4
	3	**Area per floor In sqmt**	993	618	3390	3126	2070	5865
	4	**Type of office : regular (daytime 12 hrs) BPO >16 hrs of occupancy**	daytime	daytime	daytime	BPO	BPO	day-time
	5	**% Core area :** area of lifts, staircase , toilets and other services in one floor	21%	18%	29%	22%	14%	19%
		Ranking : lower core area is ranked 1	3	2	4	3	1	2
	6	**FTF in mts :** Floor to floor clear height	2.85	2.85	3.05	3.5	2.85	4.75
		Ranking : More FTF is Rank 1 as it is good for Daylighting	4	4	3	2	4	1
	7	**DOF in mts :** Depth Of the Floor from window to window line	20	13.65	23	20	50	74.32
		Ranking : lower for better daylighting ranked 1	2	1	3	2	4	4
Envelope level	1	**U - factor** - the thermal conductance for envelope materials. Any sunshade or fins the U-Factor for fenestration is reduced as per shading co-efficient ratios (ECBC, 2017)						
		Ranking : Lowest U -factor for enevelope materials gets Rank 1	4	4	3	2	3	1
	a	**Wall U- factor**	0.68	0.78	0.63	0.63	0.63	0.57
	b	**Roof U -factor**	0.54	0.48	0.48	0.32	0.32	0.3
	c	**Glazing U- factor**	5.4	5.8	5.1	2.3	5.1	2.1
	2	**SHGC** - Solar Heat Gain Coefficient for glazing	0.53	0.72	0.53	0.32	0.53	0.3
		Raniking : Low SHGC gets rank 1	3	4	3	2	3	1
	3	**Window head height in mts -** Height of the window from	2.1	2.85	3.05	3.15	2.3	4.75
		Ranking : More window height increase daylighting	4	3	2	2	4	1
	4	**% EA to TBA :**Percentage of Envelope area (EA) to the Total	46%	62%	40%	56%	33%	25%
		Ranking : lower % of EA : TBA , the envelope area is less its	4	4	3	3	1	1
	5	**% GFA to OFA** :Percentage of Glazing per floor (GFA)to	16%	42%	26%	33%	13%	12%
		Ranking : lower % GFA: OFA is betters thermal performance as it take less thermal radiation thro glazing	2	4	3	3	1	1
	6	**WWR %** Window to wall ratio for total façade area	18%	72%	75%	70%	34%	58%
		Ranking : Low WWR better is the performance	1	3	4	3	1	2
	7	**WWR for each orientation -** Window to wall ratio on each						
		Ranking : North is prefereced and S, W & East unfavaroble , Ranking as per high on North and low on other orientation	2	4	3	1	2	1
	a	WWR % North	12%	74%	49%	87%	40%	94%
	b	WWR % South	47%	70%	49%	87%	40%	36%
	c	WWR %East	11%	71%	47%	15%	36%	49%
	d	WWR %West	11%	71%	47%	15%	36%	49%
		TOTAL POINTS	**29**	**33**	**31**	**23**	**24**	**15**
		Ranking : total of all the ranks , lowered is ranked 1	**3rd**	**4th**	**3rd**	**2nd**	**2nd**	**Ist**
Day light	1	**UDI % % 100 - 2000 lux** : Manual method as per ECBC (% 100 - 2000 lux : Manual method as per ECBC 17 the window head height for each orientation and multiply by daylight extent factor(DEF) as mentioned in ECBC (ECBC, 2017) for warm and humid climate Daylight extent factor(DEF) for warm and humid climate	41%	61%	47%	73%	10%	63%
		Ranking : Higher % of UDI is ranked 1	3	2	3	1	5	2
	2	**VT : Visual tranmittance of the glass**	0.65	0.48	0.48	0.35	0.48	0.3
		Ranking : More VT better is the view from the glass hence	1	1	1	2	1	4
ENERGY	1	**Age of HAVC : age of the equipment**	20	18	5	7	8	7
		Ranking : new equipment performace better than old , low age is ranked 1	4	4	1	2	3	2
	2	**EUI kh/m2 yr/floor .** Energy Use for one floor – daytime and BPO moreover since data was not uniform for all the samples an average is taken	301	446	130	443	554	172
		Ranking : LOW EUI is the best energy performance hence ranked 1	5	6	1	4	3	2
		TOTAL POINTS - Energy & daylight	**13**	**13**	**6**	**9**	**12**	**10**
		Ranking : total of all the ranks , best is 1st and poorest is 6th	**4th**	**4th**	**Ist**	**2nd**	**3rd**	**2nd**

Fig. 5 Table of site measurement data for the sample case buildings and their performance ranking

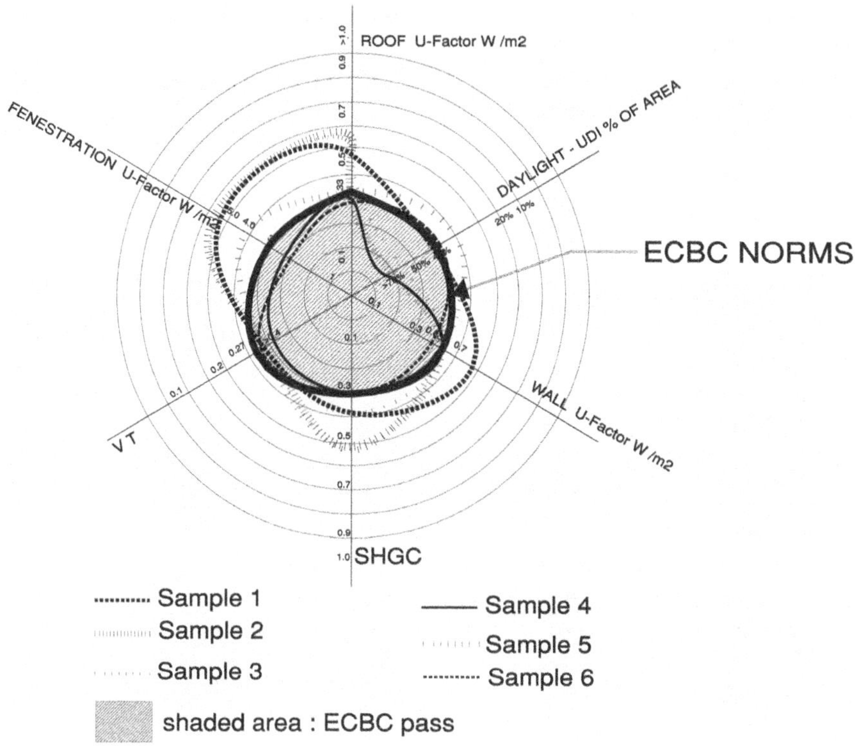

Fig. 6 ECBC 17 compliance for mandatory provisions, black line and hatched is the ECBC norm pass zone. Only samples 4 and 6 pass the ECBC norms

2.3.2 Building Envelope Trade-Off Method

The building envelope complies with the code if the energy performance factor (EPF) of the proposed building is less than the EPF of the standard building, where the standard building exactly complies with the prescriptive requirements of building envelope as in Table 1.

$$EPF\ total = EPF\ roof + EPF\ wall + EPF\ fenestrations \quad (1)$$

where EPF = Energy Performance Factors.

2.3.3 Whole Building Performance Method

Energy performance index (EPI) in KWh/sqmt / year is taken for the total usable office floor area excluding the core and utilities area, the bench mark for warm and humid zone for daytime office space with more than 50% of the floor area as air

Table 1 ECBC 2017 compliance in all the methods comparing with ECBC norms as base and case samples pass in each method

	Factors	ECBC norms	1	2	3	4	5	6
	ZONE—warm and humid zone	Building Envelope as per ECBC - 17						
	Time of completion	Core and shell	1999	2000	2003	2010	2012	2013
	SIZE—Total area in sqmt—(1) Large Office > 30,000 m. (2) Medium Office 10,000–30,000 m^2 (3) Small Office < 10,000 m^2		10,290	5830	42,124	43,673	24,675	49,962
1	*Mandatory provisions*							
1	Roof U-Factor W/m^2	<0.33	0.54	0.48	0.48	0.32	0.32	0.3
2	Wall U-Factor W/m^2	<0.63	0.68	0.68	0.63	0.63	0.63	0.57
3	Fenestration U-Factor W/m^2	<3.0	3.8	5.12	4.2	2.3	4.2	2.1
4	Visual transmittance—VT (1.0–0.1)	>0.27	0.68	0.39	0.39	0.48	0.39	0.3
5	UDI %—90% daylight time for the usable office space to achieve luminance level between 100 lx and 2000 lx for the minimum for 40% of floor area (%)	>40	41	61	47	73	10	63
6	SHGC	<0.3	0.53	0.39	0.53	0.3	0.53	0.3
	Pass mandatory provision					Pass		Pass
2	*Additional for prescriptive requirements*							
7	WWR % for warm and humid zone (%)	<40	26	72	82	70	34	58
	Pass prescriptive requirements		Pass				Pass	
3	*Building envelope trade-off method*							
	EPF—envelope performance factor	1	1.13	1.32	1.43	1	1.12	0.72

(continued)

Table 1 (continued)

	Factors	ECBC norms	1	2	3	4	5	6
4	*Whole building performance method*							
	EPI—energy performance index $KWh/m^2/yrs$	1	1.65	2.45	0.71	0.98	1.22	0.94
	Pass ECBC compliance					Pass		Pass

conditioned space is 182 KWh/sqmt/yr and occupancy with more than 12 hours per day like the BPO offices it is 452 KWh/sqmt/yr [11].

$$\mathrm{EPI} = \frac{\text{Annual Energy consumption in kwh}}{\text{total built} - \text{up area (excluding unconditioned spaces)}}$$

$$\text{EPI Ratio} = \frac{\text{EPI of proposed building}}{\text{EPI of standard building}} = \text{or} < 1$$

Observation

- Samples 4 and 6 pass the mandatory requirements; however, they fail in perspective requirement of WWR <40%.
- Samples 1 and 5 pass the perspective requirement of WWR <40% but fail the mandatory provisions.
- Only samples 4 and 6 pass the ECBC compliance for warm and humid zone in shell and core method.

3 Simulation Method

From the site data measurements, a BIM model is created in Autodesk Revit 2019. Insight is a plug-in for Revit to analyze the sample building for three major envelope performances such as average annual solar radiation, annual daylighting and space-cooling load density (CLD).

3.1 The Solar Radiation

The Insight plug-in generates data of solar insolation for annual average envelope insolation (AAEI) for analysis period of one full year and analysis time of 9am–

6 pm. Annual average solar insolation per floor area (AAIFA) is derived from formula below, Eq. 2 The same is repeated for the façade area to get annual average façade insolation (AAFI) in Wh/m^2.

Solar insolation/floor area

$$\text{Wh/m}^2\,(\mathbf{AAIFA}) = \frac{\text{AAEI Wh/m}^2 \times \text{Envelope area m2}}{\text{Total built} - \text{up area in m}^2} \quad (2)$$

Observation

As in Figs. 7 and 10.

- Sample 1—with recessed windows performed better.
- Sample 2—as the difference in AAFI and AAEI is very low, the façade receives almost equal solar insolation as the envelope.
- Sample 3 has more façade area for its cruciform shape, but envelope area is more than the total floor area; hence AAIFA is more.

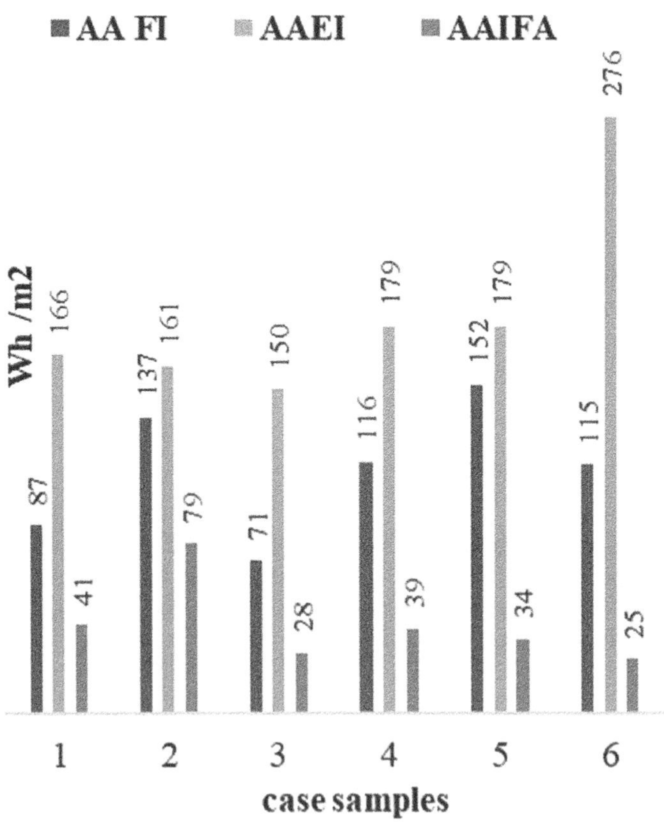

Fig. 7 Bar chart comparison of sample for AAFI, AAEI and AAIFA in Wh/m^2

- Sample 4—with linear rectangular shape and longer side facings, north and south perform better.
- Samples 5 and 6 have large floor area and DOF with lower AAIFA.
- Sample 6—with rectangular shape and corner spline fillet has very high AAFI and very low AAIFA.

3.2 *Daylight Analysis*

Annual daylighting in Revit performed within Insight plug-in called Lighting Analysis for Revit. Spatial daylight autonomy (SDA) is the % of floor area that received 300 lx for > 50% of the annual daylight time of 3650 h, and annual sun exposure (ASE) is % floor area that receives more than 1000 lx > 250 h of 3650 h.

Simulation Based on the Following Input Assumptions

- Blinds, shades or internal partitions excluded from the model.
- Default surface reflectance of 80% for ceilings, 20% for floors and 50% for walls.
- Analysis plane is at 0.8 m.
- Analysis grid is 0.6 × 0.6 m.
- 8am to 6 pm—total of 3650 h.
- Sky Component: as per annual weather data for Chennai.

Observation
As in Figs. 8 and 10.

- SDA of > 40% and ASE of <20% is a good daylighting performances.
- Sample 5 has the lowest SDA with only 4% as the window height is smaller and more depth of floor (DOF).

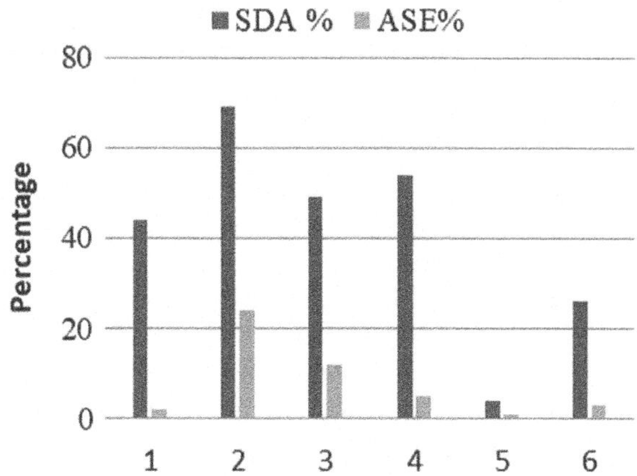

Fig. 8 Annual daylight performances for % floor area of the case samples

- Sample 6 with more DOF but the floor-to-floor height is 4.75mts and fully glazed and hence achieves SDA of 26%.
- Sample 1 recessed window with SDA of 44%.
- Sample 4 with major orientation glazed facade to north and south and double glazing has the best SDA of 54% and low ASE < 2%.
- Sample 2 fully glazed and less DOF has SDA of 69% with visual discomfort as ASE > 20%.

3.3 Energy Simulations

As cooling load is the major energy use for office building in warm and humid climate like Chennai, cooling load for one typical floor for all the samples was taken to compare the results. In order to compare the cooling load density (CLD) for the entire sample, the following fixed data input assumptions are taken to simulate the CLD in Wh/m^2.

- Office space planning: Open office plan
- Occupancy rate: 13.5 persons per sqmt
- Lighting load: 1 W/ sqmt
- Power load: 2 W/sqmt
- Cooling set point: 23.3 °C
- Analysis area: Air-conditioned space in sqmt
- HVAC type: VAV single duct system
- Occupancy hours: 9 am–6 pm (daytime)
- Supply air temperature: 12 °C
- Relative humidity: 50%
- Latent heat gain per person: 58 W
- Sensible heat gain per person: 73 W

The space selected for cooling load simulation is the sixth floor only for all the case study as it is above the tree foliage level, obstruction from the surrounding built environment, and it is sand witched by top and bottom air-conditioned spaces.

Observations
As in Figs. 9 and 10.

- The CLD for rectangle shape with WWR <70% sample 2 is very high > 100 W/ m2 and EUI of > 250 KWh/ m^2 as façade design is inefficient.
- More CLD for samples 2 and 3 with more glazing area, single glazing and less DOF as the CL component by windows > 70% and EUI > 150 $KWh/m^2/yrs$.
- Samples 5 and 6 with more DOF which has the CL components by windows are <10% and other (lighting, power, people, equipment and misc.) components more than 90%; hence, the EUI is <100 $KWh/m^2/yrs$.

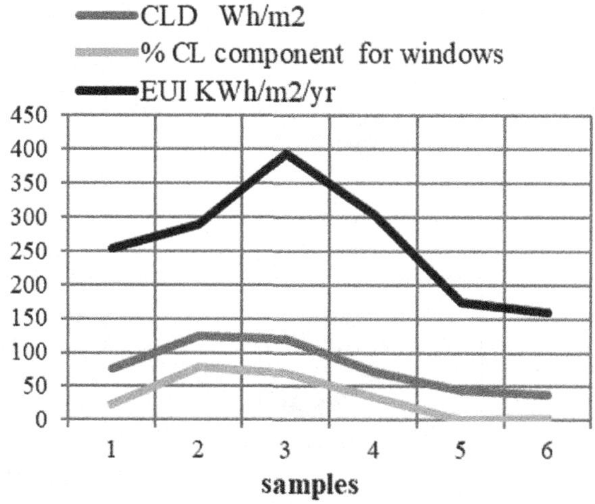

Fig. 9 Energy simulation data for the case samples

- Low CLD for sample 1 with recessed windows as more than 25% cooling load component by walls.
- Samples 6 and 4 performed the best energy-efficient envelope with double glazing and orientation to north and south and core areas located on east and west.

4 Comparison of Different Methods

The performance ranking method is a simple way to collect and organize data, but the variation in data between the samples like occupancy rate, occupancy hours and type of space planning layout, electrical consumption, floor number, context and others makes the ranking inaccurate. The ECBC 17 compliance method is also very simple numeric method for existing and new building data with just approval drawings and electricity bills. The simulation method is the best tool to design, study, analyze, identify and understand envelope energy efficiency. Simulation method is accurate than all numeric methods, but it gives too much data to select and understand what is required for the design objective. Hence, it is better to have fixed data outputs like SDA, ASE and CDL with input assumption data as mentioned above to simulate the required results. The simulation method is easy to create more design options and analysis for thermal, daylight and energy use intensity simultaneously and design energy-efficient envelope than the numeric methods.

sample	Thermal Analysis Wh/m2	Daylight Analysis 6th floor	Cooling load 6th floor & EUI total building	Cooling load Components
Case 1	 AAFI = 87 AAEI = 166 AAEFA = 41	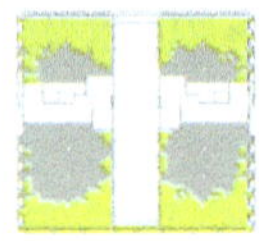 SDA- 44% ASE - 2%	 CL : 75 EUI = 216	 window 24%, walls 25 % & lighting and other 51%
Case 2	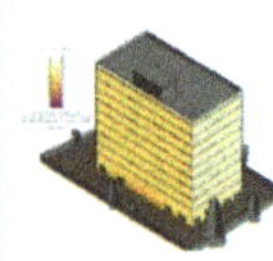 AAFI = 137 AAEI = 161 AAEFA= 79	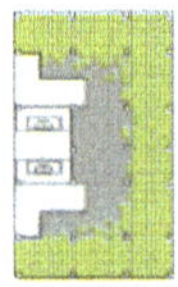 SDA- 69% ASE - 24%	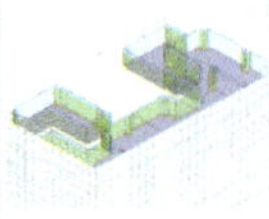 CL : 154 EUI = 256	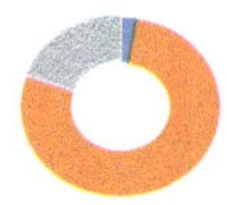 window 78%, walls 2 % & lighting and other 20%
Case 3	 AAFI = 71 AAEI = 150 AAEFA = 28	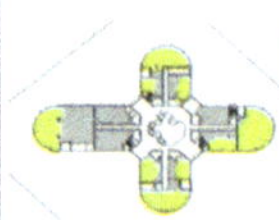 SDA- 49 % & ASE - 12%	 CD : EUI = 209	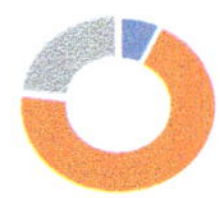 window 70%, walls 7 % & lighting and other 23%
Case 4	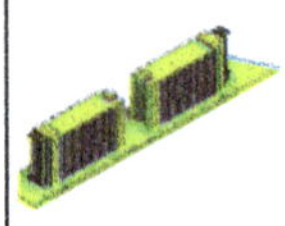 AAFI = 116 AAEI = 179 AAEFA= 39	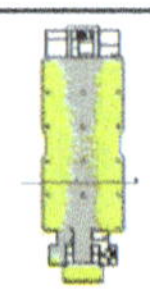 SDA- 54 % & ASE - 5%	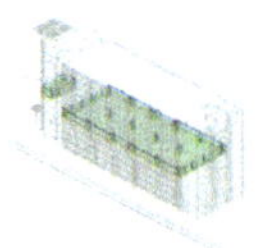 CL : 72 EUI = 152	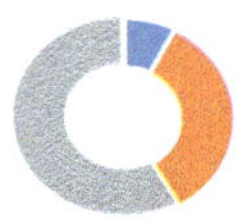 window 34%, walls 8 % & lighting and other 58%
Case 5	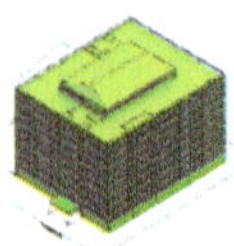 AAFI = 152 AAEI = 179 AAEFA = 34	 SDA- 4 % & ASE - 0.5%	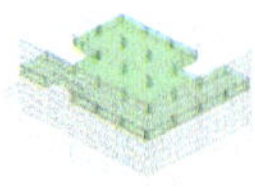 CL : 154 EUI = 269	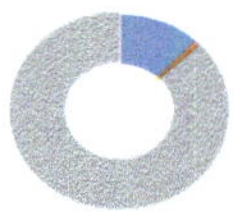 window 1 %, walls 12 % & lighting and other 87 %
Case 6	 AAFI = 115 AAEI = 276 AAEFA = 25	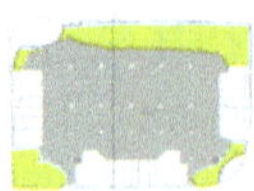 SDA- 26 % & ASE - 3 %	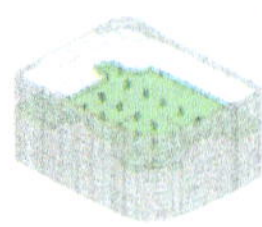 CL : 38 EUI = 88	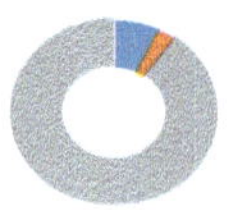 window 3 %, walls 7 % & lighting and other 90 %

Fig. 10 Simulation data with images of the samples for solar insolation, daylighting, CLD and cooling load components

5 The 13 Performance Factors for Envelop Energy-Efficient Design (EEED)

Based on the three types of data analysis, the most important EEED factors are:

EEED factor	Results and discussion	References
Building shape	Cruciform, curved or linear shapes performed better than the rectangular shape	Figures 4 and 5
Orientation of the glazing	Avoiding glazing on the south, west and east like samples 6 and 4 Location of the core/service area in these orientations makes the envelope perform better	Figures 5 and 10
Depth of the floor (DOF)	The samples 1, 2, 3 and 4 had DOF <25mts and all case had better daylighting	Figures 5 and 10
Floor-to-floor height (FTF)	With increased FTF, the daylighting performance is better. As in case 6 with 4.75 mts	Figures 5, 8, and 10
U-Factor	Lower U-Factor for envelope building material lower than ECBC 17 norms performs better	Table 1.
Glazing type (SHGC) and VLT value	SHGC < 0.3 and VLT > 0.27 is a good glazing type for better performance	Table 1. Figures 5 and 10
WWR % total and each orientation	WWR is an important factor for passive solar energy-efficient design. However, with increased WWR for fully glazed office building, proper shading device and sensible WWR % on each orientation to be incorporated.	Table 1. Figure 10
	As per ECBC standard WWR %, less than 40% but case 5 with less than 40% has the maximum cooling load, as the DOF is more. WWR < than 40% in north and south direction performed better in energy and daylight	Table 1
% ratio of glazing area per office: floor area (GFA: OFA)	As the glazing area per office area influences the cooling load density GFA < 25% is best for daylighting and energy.	Figures 5 and 10
% ratio of envelope area: total built-up area (FDA: TBA)	When < 33% the thermal performance better like in samples 5 and 6, the AA envelope isolation per floor area is very less compared to AAFI and AAEI	Figures 5 and 7
Integrated shading devices	Shading device will be required for unfavorable orientation like south, southwest and west facades in order to reduce the thermal load for this climatic context	Figures 5 and 10

(continued)

(continued)

HVAC equipment type	Better equipment like 5-star rating of BEE [12] is recommended as the return on investment is better than passive solar envelope design for operation and maintenance	
% of HVAC load to total energy load per floor	Cooling load is more than 60% of the total load	Figures 5, 9, and 10
EPF envelope performance (per ECBC)	A very important factor for energy efficiency. This factor evaluates all the physical elements of the building envelope	Figure 4 and Table 1.

6 Conclusion

- With WWR < 40%, DOF < 30mts, recessed windows with low U-factor for walls, roof, fenestration and low SHGC are an EEED and will pass the ECBC 17 compliances. As the investment for the envelope is low, this envelope design is energy efficient and affordable.
- With WWR > 70% as fully glazed building with %EA: TBA < 30%, %FGA: OFA < 25% with low SHGC will be an EEED.
- With WWR > 70% and the % FDA: TBA > 30% and % FGA: OFA > 25%, an integrated shading device or energy generating transparent solar panel (coated in glazing) will be required for an EEED.
- EPI is the most important EEED factor with just one value in KWh/m^2; however, it is inaccurate method with numeric data as per ECBC 17 for a new design; hence today, the energy simulation modeling is the simplest and accepted method.
- Installing an efficient HAVC equipment or BEE 5-star rating equipment is simpler than passive solar design as it is good return on investment, and managing and monitoring is simple and accurate.
- Simulation method has an advantage of checking the building performance and giving more options of envelope design, building elements, equipment and other data to compare and analyze the combination of different components of the energy performance models.
- The office building promoted by private developer which has multi tenants like the case samples do not intent to design for energy efficiency, hence if the private developer are given incentive like more floor space index than the standard permissible area for better star rating as per ECBC could make these buildings energy efficient.

References

1. Planning commission report (2012) Twelfth Five-Year Plan, volume-1. New Delhi, Government of India
2. National Mission for Enhanced Energy Efficiency (NMEEE)—Annual Report (2015–2016). Bureau of Energy Efficiency, New Delhi, August 2016
3. Jones Lang LaSalle (JLL) (2017) Nine Indian Cities in JLL's Latest 'Global 300' Rankings, Dec newsletter, Jones Lang LaSalle Incorporated, India, December
4. Fraunhofer Institute for Solar Energy Systems(FISES) (2017) Energy efficient buildings as central part of integrated resource management in Asian cities, Freiburg, 11th July
5. The National Action Plan on Climate Change (NAPCC) Annual report (2007–2008), Bureau of Energy Efficiency, New Delhi, August
6. India Brand Equity Foundation (IBEF) (2016) Indian real estate industry analysis report, New Delhi. December
7. Times of India Article (2012) http://article.timesofindia.indiatimes.com/2012-02-13
8. Report: Building Energy-Smart Cities is Fastest, Easiest Way to Address Indian Energy Crisis —Natural Resources Defense Council (NRDC) and the Administrative Staff College of India (ASCI), Hyderabad, October 30, 2012
9. IBN news report—ECBC for commercial building mandatory for eight states from 2012 https://ibnlive.in.com/620445.html., March 23
10. Energy Conservation Building Code 2017, Bureau of Energy Efficiency Published by Bureau of Energy Efficiency New Delhi, India, June 2017
11. BEE (2017) Energy benchmarks for Commercial Building. https://beeindia.gov.in/sites/default/files/Flyer_22nd%20Jan.pdf, January 20, 2018
12. BEE (2009) Details of the scheme for rating of office buildings Annexure 4 of BEE report February, 2009. https://beeindia.gov.in/sites/default/files/BEE%20Star%20Rating%20for%20existing%20Office%20Buildings.pdf

Optimization of Building Envelope Towards Energy-Efficient Design

G. Sudha

Abstract Buildings and construction account for more than 35% of global final energy use and nearly 40% of energy-related CO_2 emissions (Global status report 2017, International Energy Agency for the Global Alliance for Buildings and Construction). Building's envelope contributes significantly to energy consumption, especially in office buildings which use extensive glazed facades. Building envelope largely includes the various components of external façade such as walls, roof, windows and shading elements. The design of building envelope affects visual and thermal comfort in the adjoining spaces. Failure to meet the thermal and visual comfort requirements in indoor spaces results in occupants depending on mechanical conditioning systems and electrical lighting systems which increases the energy consumption in building. Hence, optimizing the envelope system will contribute largely to the energy savings in a building. This study aims at quantifying the contribution of various envelope components towards the energy optimization of an office building. Building envelope designed for an office occupancy in a warm humid climatic zone has been analysed for its energy performance in terms of various components which include thermal properties of walls, roof and glazing, window-to-wall ratio and shading elements. Analysis has been done by simulating the envelope optimized building design against the base case as prescribed by Energy Conservation Building Code, using an environmental analysis software tool. The results establish the impact of various envelope components on the energy consumption of the building. The findings emphasize the provision of an appropriate shading device as the highest contributor towards achieving the goal of energy efficiency by reducing the solar heat gain. Hence, careful consideration must be given in designing the shading elements in warm humid climatic zone as it can significantly change the performance of other components of an integrated building, namely HVAC and artificial lighting system, resulting in an energy-efficient design.

G. Sudha (✉)
School of Architecture and Interior Design, SRM Institute of Science and Technology, Kattankulathur, Chengalpattu, Tamilnadu, India
e-mail: archisudha@gmail.com

K. Thirumaran et al. (eds.), *Sustainable Urban Architecture*,
Lecture Notes in Civil Engineering 114,
https://doi.org/10.1007/978-981-15-9585-1_12

Keywords Building envelope · Envelope optimization · Energy efficiency · Energy optimization · Shading device · Window–wall ratio

1 Introduction

Buildings' energy consumption represented about 14% of total delivered energy consumption in India in 2015. EIA's International Energy Outlook 2017 (IEO2017) projects that among all regions of the world, the fastest growth in buildings' energy consumption through 2040 will occur in India. In the IEO2017 Reference Case, delivered energy consumption for residential and commercial buildings in India is expected to increase by an average of 2.7% per year between 2015 and 2040, more than twice the global average increase. Most of this growth is the result of increased electricity and natural gas use (because of greater access to these energy sources) and the increased use of appliances and energy-intensive equipment [1].

EIA projects the electricity share of India's total commercial energy consumption to continue increasing, from 59% in 2015 to 65% in 2040, displacing some coal consumption. EIA projects that total delivered commercial sector energy use in India will increase by an average of 3.4% per year—again, the fastest growth rate among IEO regions. Global residential and commercial sector energy consumption is also on an increase as shown in Fig. 1.

The two factors which dictate the energy consumption of a building are the building's active systems (HVAC, lighting, equipment, etc.) and passive architectural design strategies (form, orientation, envelope design, construction, etc.). Among all the above-mentioned factors, the building envelope contributes significantly to the total energy consumption which has been established in various studies. A study in Hong Kong [2] has found that the building envelope design

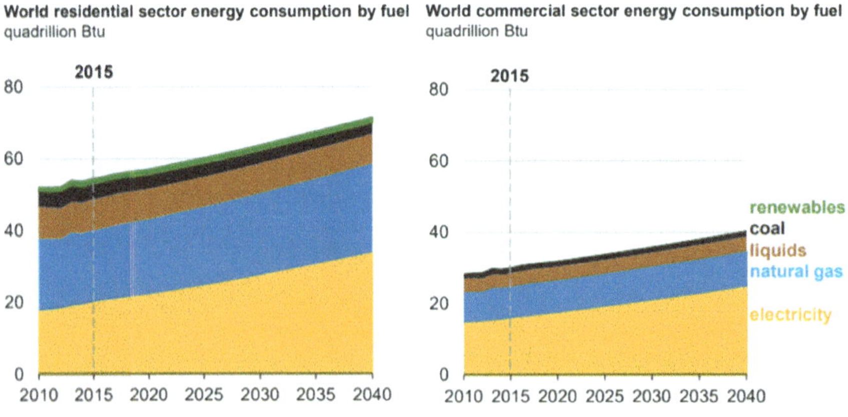

Fig. 1 World residential and commercial sector energy consumption [1]

accounted for 36% of the peak cooling loads in office buildings. Another study [3] has simulated office buildings in Abu Dhabi and found that the building envelope was responsible for 30% of the building's total cooling loads (solar, glass and fabric). Building envelope consists of elements comprising the external façade, namely opaque walls, fenestrations, roof and shading devices.

According to Brock [4], building envelope is the skin of a building which is supported by the skeleton of the building structure. It acts as a thermal barrier between the enclosed conditioned space and outside environment through which the thermal energy is transferred. By minimizing the heat transfer through the building envelope, the need for energy used in space heating and cooling can be reduced considerably.

Various studies have been carried out to analyse the impact of different building envelope components such as walls, roof, WWR, glazing and shading devices separately on the energy performance of a building for various climatic conditions. A study has concluded that the heating and cooling loads of buildings in the Gaza Strip are gradually reduced, as the wall U value is lowered [5]. Due to the high air temperature and abundance solar radiation characterizing the climate of Jeddah, the thermal characteristic of the glazing type such as shading coefficient is the dominant factor in reducing cooling energy loads in relation to solar gains [6].

This study aims at identifying the element that contributes the most towards an energy-efficient building envelope under warm humid climatic conditions of Chennai.

2 Energy Conservation Building Code (ECBC)

ECBC was developed by Ministry of Power and Bureau of Energy Efficiency (BEE) to establish minimum energy performance standards for commercial buildings in India, having a connected load of 100 kW or greater or a contract demand of 120 kVA or greater. The buildings falling under the scope of this code have to comply with the mandatory requirements and any of the compliance path—prescriptive method or whole building performance method prescribed by the code.

Prescriptive method defines the minimum or maximum compliance requirements for various envelope components and energy-consuming systems of the proposed building in addition to the mandatory requirements. Whole building performance method requires the estimated annual energy use of the proposed design to be less than the standard design (whose enclosure elements and energy-consuming systems are designed according to the reference requirements mentioned for the same in the code) in addition to meeting the mandatory requirements.

In addition to the above two methods, the code also describes building envelope trade-off method which is a system-based approach where the thermal performance of individual envelope components can be reduced if compensated by higher efficiency in other building components. This study is based on the whole building performance method where the estimated annual energy use of the proposed design has been compared with the standard base case.

3 Research Aim, Scope and Method

The aim of this study is to identify the various elements of building envelope that contribute to the building's energy consumption, compare the energy performance of each element and identify the element that has the most significant impact on the energy performance of a multi-storied office building in the warm humid climatic zone of Chennai.

The scope of this study is limited to analysing the energy performance of the building in terms of annual energy consumption and energy performance metrics—energy use intensity (EUI), which is a result of the envelope's ability to reduce the heat gain through its various elements. Architectural passive design strategies such as form and orientation have been integrated while creating the model, but energy performance analysis of these architectural design parameters and spatial configuration analysis do not form a part of this study.

A hypothetical stilt +13 floors study model has been created in accordance with the prevailing development rules and regulations laid down by the Chennai Metropolitan Development Authority. The study has been developed through dynamic simulations to evaluate the energy performance of identified building envelope elements. A base case has been modelled by specifying the thermal properties of various elements as stipulated in ECBC, and results are compared with the proposed design case which has thermal properties of building envelope elements exceeding beyond the values stipulated by ECBC and additional elements not specified in the building code (shading devices).

The use of building simulation software for buildings' dynamic analysis is a necessary and well-established procedure to study effective building energy performance given real climate considerations [7]. The selected tool, Sefaira, is a building simulation tool capable of performing dynamic thermal simulation of buildings. It is a collaborative, cloud-based software that uses industry standard analysis engines such as Energy Plus, Radiance and Daysim [8]. It not only allows tracking various performance metrics such as energy use intensity (EUI), daylight metrics and thermal comfort metrics but also provides data on HVAC sizing, renewable system sizing and the operating costs. Sefaira conducts full hourly annual simulations throughout a typical year based on the weather (EnergyPlus Weather) files uploaded by the user.

4 Climatic Analysis of Chennai

Chennai, located along the eastern coast of Indian subcontinent, falls under warm humid climate according to the ECBC climatic classification. EnergyPlus Weather data files for Chennai have been used in the simulation. Climatic analysis of Chennai has been done based on the data generated from Climate Consultant software which uses the EPW data files.

The climate of Chennai is characterized predominantly by slightly high air temperatures and high relative humidity levels. The model used for simulation has been created integrating the passive strategies arrived based on the climatic conditions of Chennai. Sun shading charts (Figs. 2 and 3) and the overheated period of Chennai (Fig. 4) indicate strongly the requirement for proper shading devices almost all round the year.

5 Base Case—Model and Input Parameters

The base case model has been constructed based on the prevailing development rules and regulations laid down by the Chennai Metropolitan Development Authority. The model consists of S+13 floors with all specifications assumed as outlined in Table 1. The model and zones for simulation are fixed for both base case and proposed design case as spatial configuration analysis is not a part of scope of this study.

The parameters used in simulation are based on the prescriptive requirements described in ECBC for an energy-efficient building envelope. The code sets the maximum prescriptive requirements for the U value and SHGC of glazing, U value of roof assembly and opaque wall assembly.

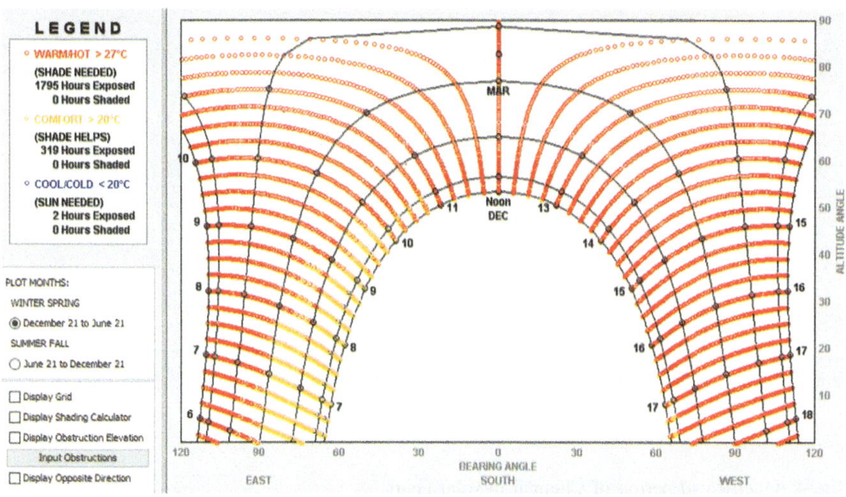

Fig. 2 Sun shading chart (January to June)—climate consultant tool

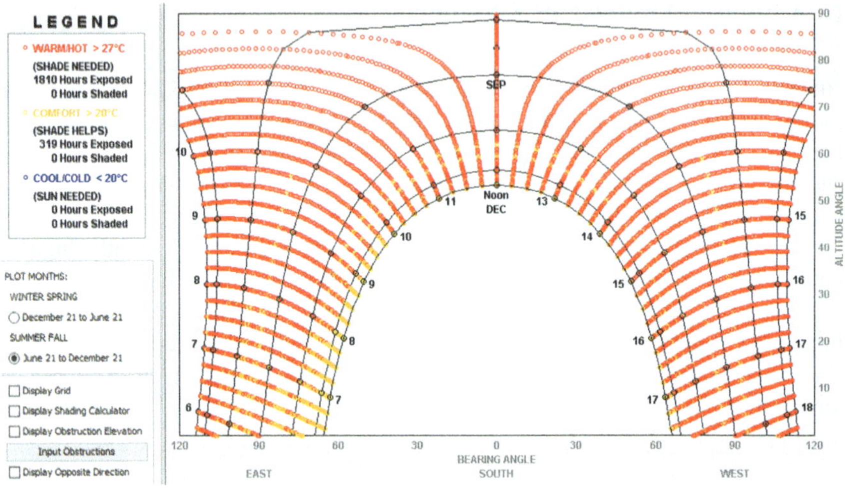

Fig. 3 Sun shading chart (July to December)—climate consultant tool

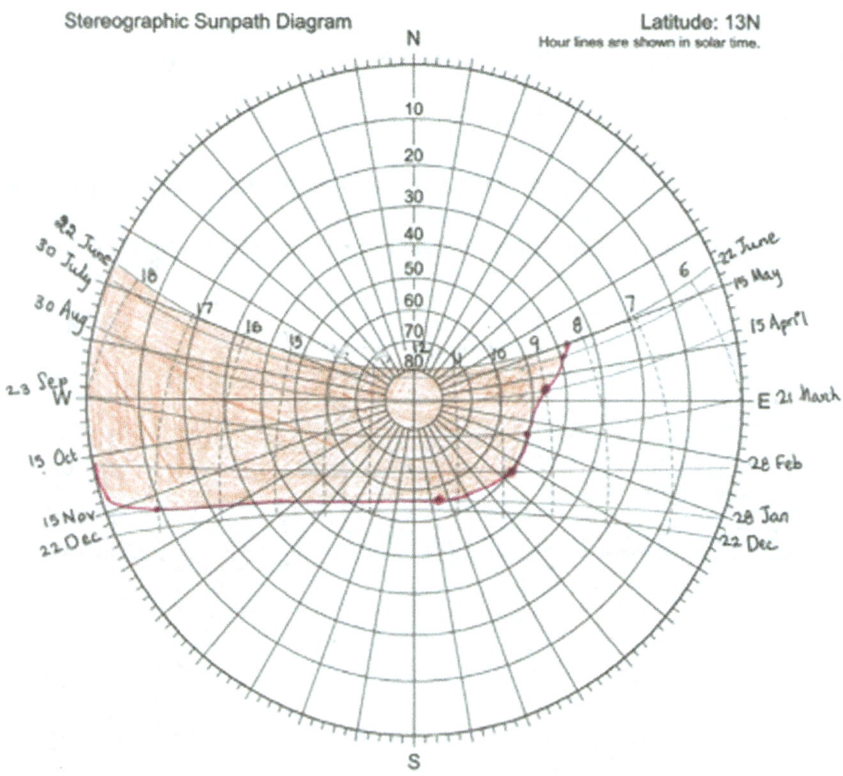

Fig. 4 Overheated period of Chennai in solar chart

Table 1 ECBC requirements

<table>
<tr><td colspan="5">ECBC building envelope requirements for warm humid climatic zone</td></tr>
<tr><td>Maximum WWR</td><td colspan="4">60%</td></tr>
<tr><td colspan="5">Facade glazing</td></tr>
<tr><td>Max U value</td><td colspan="4">3.0 W/m² K</td></tr>
<tr><td rowspan="2">Max solar heat gain coefficient (SHGC)</td><td colspan="2">North orientation</td><td colspan="2">Non-north orientation</td></tr>
<tr><td colspan="2">0.27</td><td colspan="2">0.27</td></tr>
<tr><td rowspan="2">Min visible light transmittance (VLT)</td><td>WWR 10%–30%</td><td>WWR 31%–40%</td><td>WWR 41%–50%</td><td>WWR 51%–60%</td></tr>
<tr><td>0.27</td><td>0.20</td><td>0.16</td><td>0.13</td></tr>
<tr><td colspan="5">Overall roof assembly</td></tr>
<tr><td>Max U value</td><td colspan="4">0.33 W/m² K</td></tr>
<tr><td colspan="5">Overall opaque external wall assembly</td></tr>
<tr><td>Max U value</td><td colspan="4">0.63 W/m² K</td></tr>
<tr><td colspan="5">Shading</td></tr>
<tr><td colspan="5">As per the building massing and no additional shading devices have been provided</td></tr>
<tr><td colspan="2">Air cooled chiller</td><td colspan="3">Condenser water loop</td></tr>
<tr><td>Chiller COP</td><td>3.0</td><td colspan="2">Heat rejection source</td><td>Cooling tower</td></tr>
<tr><td>Chilled water temperature—supply</td><td>7 °C</td><td colspan="2">Condenser water temperature—supply</td><td>29.4 °C</td></tr>
<tr><td>Chilled water temperature—return</td><td>12 °C</td><td colspan="2">Condenser water temperature—return</td><td>36 °C</td></tr>
<tr><td colspan="5">Space use settings</td></tr>
<tr><td>Occupant density</td><td colspan="4">10 m² per person</td></tr>
<tr><td>Equipment power density</td><td colspan="4">25 m² per person</td></tr>
<tr><td>Lighting power density</td><td colspan="4">10 m² per person</td></tr>
<tr><td>Set point temperatures</td><td colspan="4">21 °C–24 °C</td></tr>
<tr><td>Outside air rate per person</td><td colspan="4">15L/s-person</td></tr>
</table>

5.1 *Base Case Annual Energy Analysis*

Simulation is done for the base case with the prescriptive requirements described in ECBC as listed in Table 1. The base case energy simulation result (Fig. 5) shows that the cooling load contributes to 53% of energy consumption in the building (AHU cooling—43%, fans—9% and pumps—1%) and the interior lighting and equipment load contributes to 17% and 31% of the total energy consumption, respectively. This clearly gives an indication that the optimization of building envelope would contribute to the reduction in energy requirement for building cooling and interior lighting.

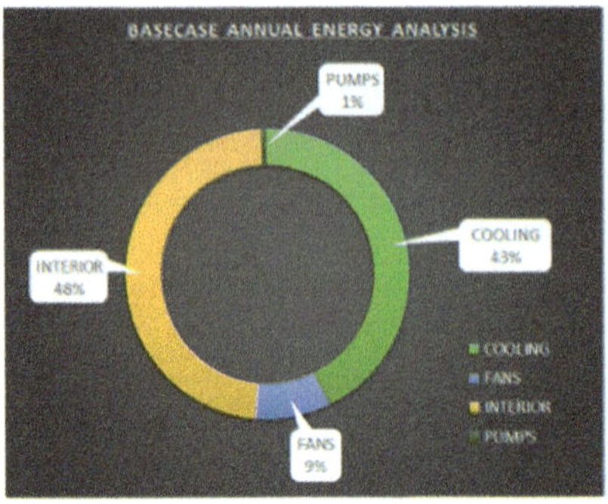

Segment wise Annual Energy (kWh per yr)		% of total
AHU	**43%**	
Air Handling Unit	1136805	43%
FANS	**9%**	
Air Handling Unit	233606	9%
PUMPS	**1%**	
Pumps	16847	1%
INTERIOR	**48%**	
Lighting	460751	17%
Equipment	818329	31%
TOTAL	**2666338**	**100%**

Fig. 5 Base case—break-up of annual energy consumption

6 Proposed Design Case

The peak heat gain contributors identified from the simulation are as follows:

- Glazing conduction and solar heat gain through fenestration
- Solar heat gain through opaque walls
- Roof gain

As a result, the following parameters have been optimized in the proposed design case in order to identify the most significantly contributing one:

- Window–wall ratio
- Glazing properties—U value and SHGC
- U value of roof assembly
- U value of opaque wall assembly
- Shading devices

The influence of each of the above parameters on the energy consumption of the building has been studied, analysed and presented below.

7 Proposed Design Case Energy Analysis

7.1 Window–Wall Ratio

In a tropical climate, buildings can have a wider range of WWR values especially when the proper type of glazing (low U value and solar heat gain coefficient) and the shading systems are employed in the facades for solar heat gain control [9]. Optimum WWR which strikes a balance between cooling energy demand and artificial lighting energy demand has to be selected.

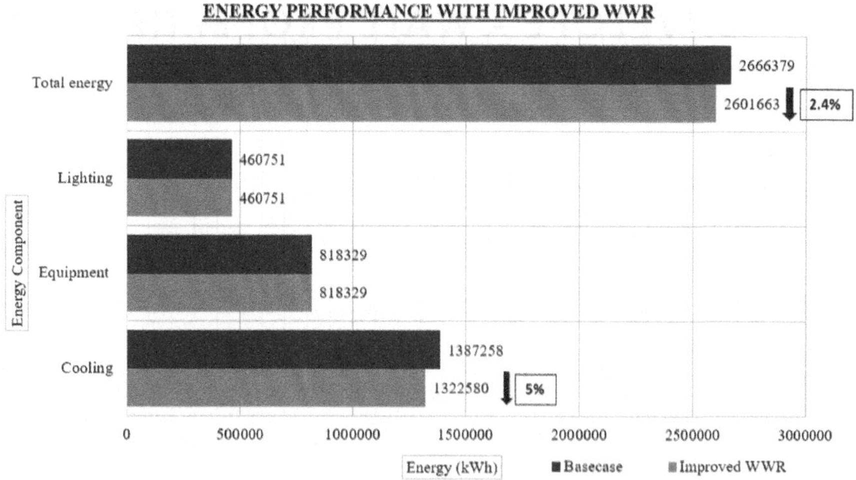

Fig. 6 Comparison between base case and proposed design case with improved WWR

Figure 6 shows that, with WWR reduced to 40%, the total annual energy consumption has dropped down by 2.4% from the base case which is mainly due to the reduction in building cooling load by 5% (AHU cooling, fans and pumps). The energy consumption by equipment and interior lighting remains unaltered.

7.2 Opaque Wall Assembly

As a part of building's envelope, walls account for a significant proportion of heat loss and gain [10]. U value or thermal transmittance is a thermal property of overall opaque wall assembly which is to be critically considered while assessing the energy performance of buildings. U value measures the amount of heat transmitted (heat gain/heat loss) through the thickness of all the elements that forms a building component such as wall or roof. It is an indicator of the insulating property of a building component. The lower the U value, the lesser is the heat transmitted through the building fabric and the better insulator it is. U value depends on the thermal conductivity and the thickness of the individual material in the composite wall/roof assembly.

Thormark (2006) reported that improving the thermal insulation of the building envelope will considerably reduce the energy required to operate the building [11]. Mishra et al. (2012) estimated the optimum thickness of wall insulation materials as a way to reduce the energy consumption of buildings in India, taking into consideration the cost of insulation and energy over the lifetime of the building [12].

Reducing the U value of opaque wall assembly from the ECBC base case value of 0.63 results in a gradual reduction of EUI as shown in Fig. 7. Hence, a composite

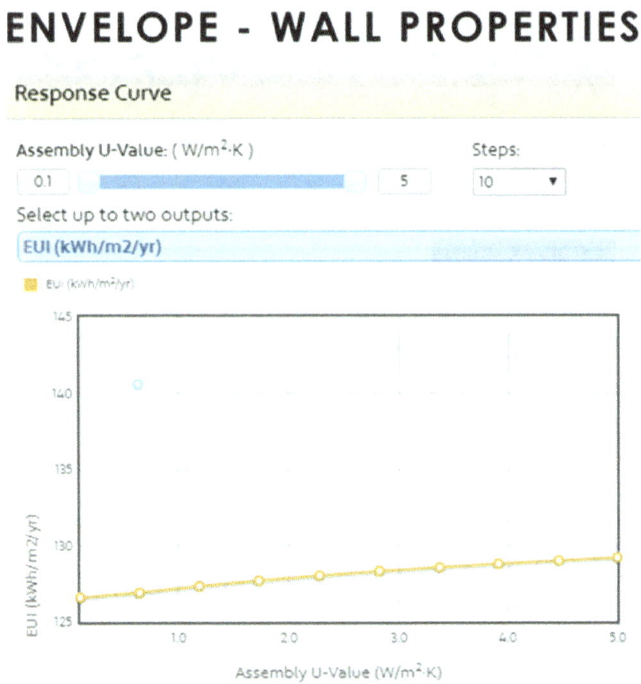

Fig. 7 Response curve showing the relationship between U value of wall assembly and EUI

wall section has been developed with 75-mm-thick rock wool insulation sandwiched between 200-mm-thick AAC blocks as the exterior layer and 100-mm-thick fly ash brick as the interior layer as shown in Fig. 8.

AAC block has a thermal conductivity of 0.18 W/m K as against a conventional clay brick of thermal conductivity 0.81 W/m K. U value of the overall wall assembly has been calculated as 0.3 W/m^2 K using assembly U factor calculator tool developed by Centre for Advanced Research in Building Science and Energy (CARBSE), CEPT University, Ahmedabad, India.

Simulation done with the revised U value of opaque wall assembly yields results as shown in Fig. 9. Total annual energy consumption reduces by 0.4% from base case which is contributed by the reduction in energy consumption by cooling equipment by 1%.

7.3 *Roof Assembly*

Roof is the most exposed building envelope element to solar radiation and hence the largest contributor to heat gain in a building (single storeyed). The thermal performance of roof is also characterized by the U value of overall roof assembly.

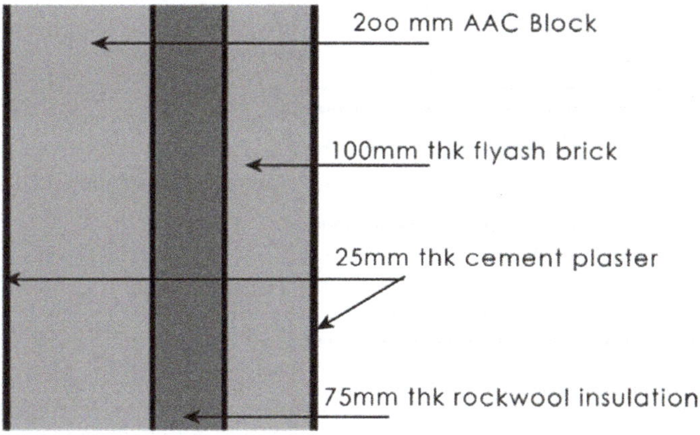

Fig. 8 Section of proposed wall assembly

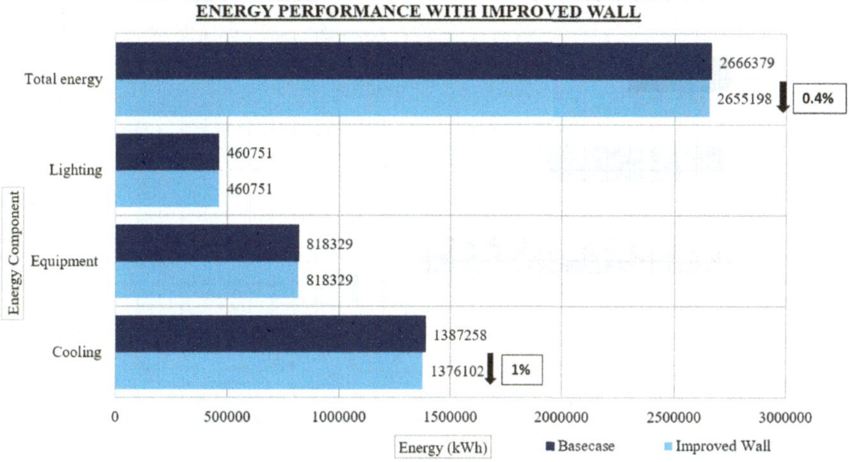

Fig. 9 Comparison between base case and proposed design case with improved U value of wall

Roof optimization has been done by providing a 75-mm-thick polyurethane foam insulation above RC slab and a layer of cool tile and mortar as the exterior finish (Fig. 10). This roof configuration reduces the U value of overall roof assembly to 0.292 W/m^2 K which has been calculated manually. Simulation done to analyse the energy consumption pattern shows reduction in the annual energy consumption by 0.4% from the base case (Fig. 11) which is mainly due to the reduction in cooling load by 1%.

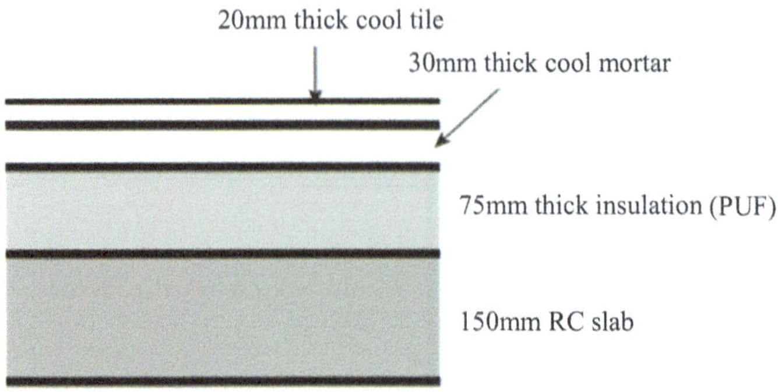

Fig. 10 Section of proposed roof assembly

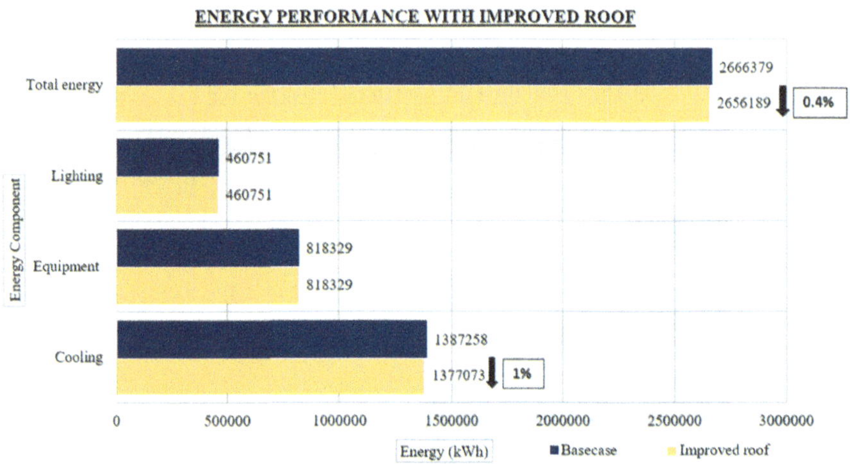

Fig. 11 Comparison between base case and proposed design case with improved U value of roof

7.4 *Façade Glazing*

Solar heat gain through glazing is a significant factor in determining the cooling load of a building. Energy performance characteristics of glazing that determines this are solar heat gain coefficient (SHGC) and U value. SHGC measures the amount of heat transmitted through the glazing as a fraction of incident solar radiation. SHGC is dimensionless ranging from 0 to 1. The lower the SHGC, the lesser is the heat gain through glass.

Another important characteristic to be considered in conjunction with SHGC is visible light transmittance (VLT) of glass. VLT is defined as the amount of light in the visible portion of spectrum that passes through glass. SHGC and VLT of glass are determined by the glazing type, number of panes and the glass coatings.

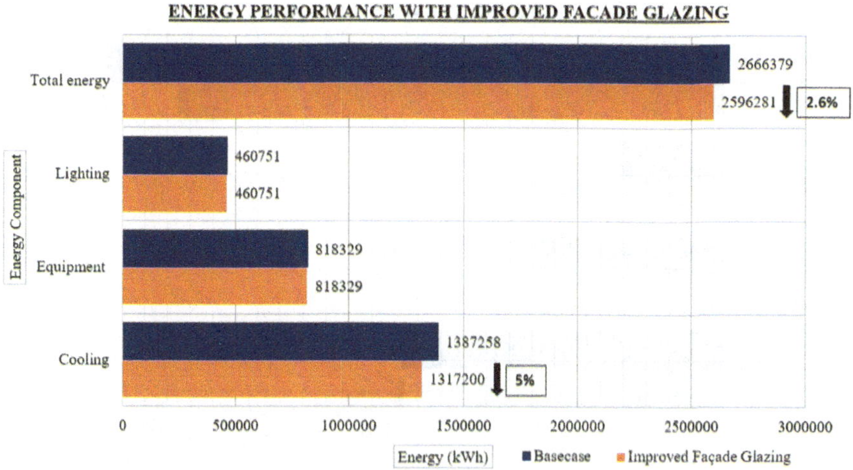

Fig. 12 Comparison between base case and proposed design case with improved glazing properties

Windows provided with various tints/coatings to reduce the solar heat gain considerably reduce its VLT also. Hence, balance between the two characteristics SHGC and VLT has to be maintained during the selection of glazing material. Low e coating in glass with emittance as low as 0.04 reflects 96% of the incident long-wave infrared radiation, thus reducing the SHGC significantly with little reduction in VLT.

The study uses a double-glazed unit with a 6-mm coated glass (coating face 2) +12-mm air gap +6-mm low e glass which has SHGC 0.27, VLT 37% and U value 1.6 W/m^2 K in the simulation. The reduction in U value and SHGC of glass has reduced the solar heat gain and conduction through glass, thus reducing the cooling load by 5% and annual energy consumption by 2.6% from the base case (Fig. 12).

7.5 *Shading*

Another way to reduce the solar heat gain through fenestration is to provide proper shading devices for these fenestrations and for the walls also. But this is seldom considered to the fullest extent in the energy codal provisions. This study aims to quantify the contribution of shading devices towards the reduction in energy use of an office building.

Fenestration without shading allows the transmission of solar radiation inside the space, and problems of glare and visual discomfort also become inevitable. Shading devices contribute to both thermal performance and daylight in a building. Hence, an integrated approach is required, taking into account the balance between these

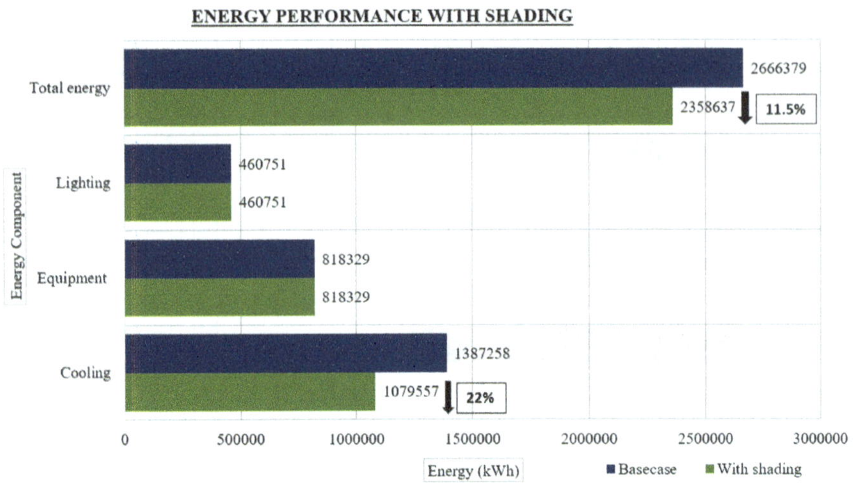

Fig. 13 Comparison between base case and proposed design case with shading

two parameters, to arrive at an optimal solution. Design of shading devices is governed by the location and orientation for which it is designed.

The final simulation has been done by providing shading for all the fenestrations and walls which reduces the heat gain by conduction through opaque walls and glazing and also the solar heat gain through glazing. This has resulted in the reduction of total annual energy consumption by 11.5% from the base case contributed by the reduction in cooling equipment load (AHU, fans and pumps) by 22% from the base case (Fig. 13). Interior lighting load and equipment load remain unaltered.

7.6 *Energy Use Intensity (EUI)*

EUI arrived from the simulation for the base case is 141 kWh/m^2/year (Fig. 14) which has reduced to

- 137 with the reduction in SHGC of façade glazing (EUI reduced by 2.6% from the base case)
- 138 with the optimization of WWR to 40% (EUI reduced by 2.4% from the base case)
- 140 with the reduction in the U value of opaque wall assembly and roof assembly (EUI reduction by 0.4% each from the base case)
- 125 with the provision of proper shading devices for the fenestration and walls (EUI reduced by 11.5% from the base case)

Efficient shading devices designed for the fenestration and walls have contributed largely to the reduction of EUI.

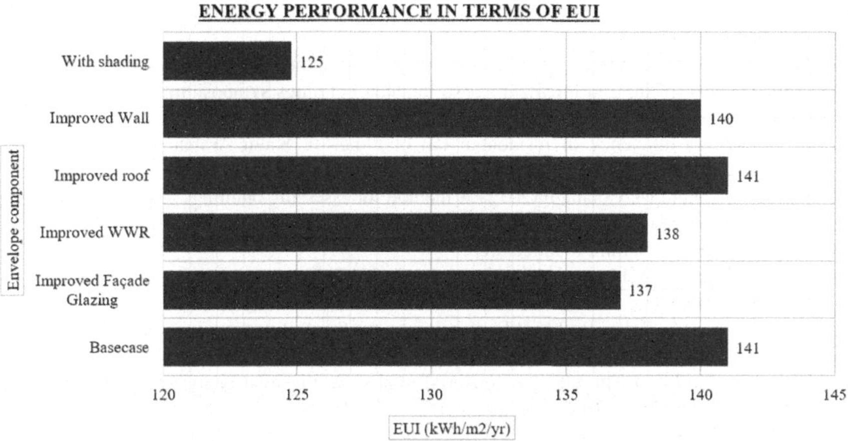

Fig. 14 Comparison between EUI of base case and proposed design case with various iterations

8 Conclusion

The study aims to explore the contribution of various envelope parameters that impact the thermal performance of an office building in the warm humid climatic region of Chennai and identify the most significant parameter. The simulation results show that altering the WWR, U value of opaque wall assembly and roof assembly, façade glazing properties and shading reduces the annual energy consumption by 2.4, 0.4, 0.4, 2.6 and 11.5%, respectively. This emphasizes the fact that optimizing WWR, glazing properties and shading have a significant reduction in the solar heat gain with shading being the largest contributor to reduction in cooling load. Thus, focusing on the architectural design parameter such as shading for both fenestrations and opaque walls is one of the most promising solutions to improve the thermal performance of a multi-storied office building in the warm humid climatic region of Chennai, which can aid in designing an energy-efficient building envelope. The impact of shading devices of each orientation on the building's energy performance could further be explored separately, in order to optimize its design and construction.

References

1. U.S. Energy Information Administration. International Energy Outlook 2017 Reference case
2. Haase M, Amato A (2006) Sustainable facade design for zero energy buildings in the tropics. In: 23rd conference on passive and low energy architecture (PLEA), Geneva, Switzerland
3. Al-Hosany NK (2002) Sustainable facade design and virtue in incarceration architecture: the case of prison buildings in Abu Dhabi, PhD Thesis, University of Newcastle

4. Brock L (2005) Designing the exterior wall: an architectural guide to the vertical envelope, 1st edn. Wiley, New Jersey
5. Muhaisena A (2015) Effect of wall thermal properties on the energy consumption of buildings in the Gaza strip. In: 2nd international sustainable buildings symposium
6. Ghabra Noura, Rodrigues Lucelia, Oldfield Philip (2017) The impact of the building envelope on the energy efficiency of residential tall buildings in Saudi Arabia. Int J Low-Carbon Technol 12(4):411–419
7. Pisello A, Goretti M, Cotana F (2012) A method for assessing buildings' energy efficiency by dynamic simulation and experimental activity. Appl Energy 97:419–429
8. https://sefaira.com/sefaira-architecture/
9. Raji B, Tenpierik MJ, van den Dobbelsteen A (2017) Early-stage design considerations for the energy-efficiency of high-rise office buildings. Sustainability 9
10. Rhee-Duverne S, Baker P (2013) Research into the thermal performance of traditional brick walls, English Heritage research report 2
11. Thormark C (2006) The effect of material choice on the total energy need and recycling potential of a building. Build Environ 41(8):1019–1026
12. Mishra S, Usmani JA, Varshney S (2012) Energy saving analysis in building walls through thermal insulation system. Int J Eng Res Appl (IJERA) 2(5):128–135

Optimization of Building's Wall Using Phase Change Material (PCM) Toward Energy Performance Improvement

C. Piraiarasi, Saravana Kannan Thangavelu, and Mhd Faizal Bin Mansur

Abstract Building sector has contributed the largest energy consumption, and most energy is consumed by existing buildings. Rate of replacing the old buildings is only around 1.0–3.0% annually. Therefore, to reduce global energy usage, enhancement of energy efficiency in existing buildings is necessary. Building Information Modelling (BIM) and analysis can improve building energy performance by providing precise strategy in retrofitting the building envelope. Currently, retrofitting is commonly done following a global trend in maximizing the energy efficiency of existing buildings. As such, this paper seeks to improve the energy performance of a building by simulating encapsulated phase change material (PCM)-enhanced cellulose insulation inside the wall that will create sustainable indoor environment for the occupants. Various solutions for geometrical and energy-based data acquisition are explored, and selection of the most appropriate BIM technology and energy analysis software (Revit and EnergyPlus) has been decided. Both softwares are interoperable via the same platform of schemes. Technical research in both engineering and energy domain sector is conducted to ensure the reliability of the results. Comprising of both geometrical and energy-based data acquisitions from the selected building, a thorough generation of building modeling was performed. Then, transfer of energy-based data into the energy analysis software and identification of the main issues related to the energy performance are determined. Finally, retrofitting of the walls using PCM modules in EnergyPlus was achieved. Results indicate that the energy consumption by using PCM-incorporated wall can be optimized up to 11.23% annually compared to a typical wall.

C. Piraiarasi (✉)
Department of Architecture, Thiagarajar College of Engineering, Madurai, Tamil Nadu 625015, India
e-mail: cparch@tce.edu

S. K. Thangavelu · M. F. B. Mansur
Swinburne University of Technology, Sarawak Campus, 93350 Kuching, Sarawak, Malaysia
e-mail: sthangavelu@swinburne.edu.my

M. F. B. Mansur
e-mail: sakaenergy@gmail.com

K. Thirumaran et al. (eds.), *Sustainable Urban Architecture*,
Lecture Notes in Civil Engineering 114,
https://doi.org/10.1007/978-981-15-9585-1_13

Keywords PCM · Retrofitting · Building envelope · BIM · Revit · EnergyPlus

1 Introduction

Energy usage in developing countries such as South East Asia, Middle East, South America and Africa is estimated to grow at 3.2% annually. Furthermore, China only took 20 years to double its energy consumption with a whopping 3.7% growth rate compared to other countries [1]. This statistic shows that energy usage is closely related to the development of a country. Hence, improving the energy efficiency will sustain the environment.

Compared to other sectors, building sector has contributed the largest energy consumption reaching 20–40% in developing countries right above industrial and transportation sectors [1]. Retrofitting strategy is a method to improve the energy performance of a building. One way to achieve this is by using the Building Information Modelling software that can analyze the energy performance of the building. This method is widely used in structural design analysis concomitant with the developing technologies of the world. In the building sector, most energy is consumed by existing buildings. Rate of replacing the old buildings with the new one is only around 1.0–3.0% annually. Therefore, to prevent the increment of global energy usage, rapid enhancement of energy efficiency in existing buildings is necessary [2]. One of the approaches for cutting down building energy consumption and enhancing indoor thermal environment is to integrate phase change material (PCM) into a building or building services system [3].

In Malaysia, BIM technology has been well accepted and implemented in Construction Company. However, the development of BIM technology is rather significantly slow. Most adoption of BIM only focused on architectural industry with only little research that involves engineers or engineering consulting services [4]. Retrofitting through BIM helps the engineers to identify the usage and improving the performance of energy and its efficiency. During the last decade, a lot of government and international organizations contributed to the energy efficiency improvements globally [2]. However, lack of data from reliable sources is one of the challenges to retrofit existing building [5]. Other factors such as climate change are also one of the uncertainties when retrofitting.

In this research, simulation is done on the building's wall to analyze the energy performance level of the current Malaysia building and retrofitting the existing wall using PCM to improve the energy usage.

2 Background

2.1 *Building Envelope*

Building envelope is the physical separator between the conditioned environments of a building including the resistance to air, water, heat, light and noise transfer. It has several functions such as structural support for the building, to control the indoor climate of the building as well as aesthetic value for interior design of the building. The main focus of this research paper is to do evaluation of the building envelope of a building.

The building envelope serves as a protective medium against these loads such as wind, rain, solar radiation, air pollution and external industrial hazards such as carbon dioxide and nitrogen gases. Building envelope plays an important role in environmental sustainability because it affects the energy efficiency of the building including the material efficiency as well as thermal comfort of the occupants. Moreover, by optimizing the building envelope, it will reduce the building resource consumption and environmental degradation [6].

2.1.1 Walls

Walls provide thermal comfort for the building's interior. It separates the outer environment from inside conditioned space of the building. Walls act as thermal resistance (R-value) that will affect the consumption of the building thermal energy. Most of the walls will be installed with thermal insulator.

2.2 *Heating Ventilating and Air-Conditioning (HVAC)*

HVAC functions as a system that controls the level of dampness and indoor temperature inside a building. Energy used to manage HVAC is estimated about 50% in a classic commercial building [7]. This implies that half of the energy usage of a building comes from the HVAC system.

However, most HVAC systems in buildings are still using the same primitive method to maintain the building systems. By implementing building energy management system (EMS) and data from occupancy nodes to the HVAC systems, electrical energy consumption can be saved up to 9.54–15.73% while thermal energy saved up to 7.59–12.85% [8].

2.3 Phase Changing Material (PCM)

Phase change materials (PCMs) are materials that can change phases when exposed to certain range of temperature. It is able to store large amount of energy because of its high latent heat of fusion while changing its phase from solid to liquid. Typically, PCM is installed in building envelopes such as walls, floors, ceilings and roofs to decrease the heat transfer rate during peak hours and maintaining comfortable ambient temperature of indoor environment. Thermal energy can be absorbed and stored in PCM through a melting process during daytime. At night, it solidifies and releases a portion of the stored energy into the surrounding environment. PCM has a number of merits including storing thermal energy during daytime and releasing it at night, particularly for space heating in the winter season; shifting electricity consumption from peak period to off-peak the electricity demand; and improving thermal comfort condition by reducing the swing of indoor air temperature. PCM can be incorporated into a solar heating system, chilled ceiling, underfloor air-conditioning, building roof, window glazing as well as the external wall of a building [3, 9, 10].

PCM can be classified into organic and inorganic materials. Organic materials are paraffin and non-paraffin; meanwhile, inorganic materials consist of hydrated salts and metals. Paraffin that has chemical formula of CH_3–(CH_2)–CH_3 is thermally and chemically stable. Paraffin also shows resistance to corrosion as well as has high latent heat. However, when used in construction of buildings, paraffin has the tendency to slip from the matrix and is flammable with low thermal conductivity despite being economically affordable. Non-paraffin PCM on the other hand has high latent heat with low supercooling and provides various choices to the user but will cost more compared to paraffin. Examples of non-paraffin are esters, glycols and fatty acids [11].

2.4 BIM Software

BIM software does not work independently to analyze the energy performance of a building. It usually needs an integration of several softwares to produce more reliable and dependent results. BIM authoring software is needed to design the 3D structure of the building into a virtual set of data, while energy analysis software will be used to simulate the energy performance of the building using the file exported from the authorized software [12].

2.4.1 Autodesk Revit

Autodesk Revit is chosen as the BIM software to design the 3D model of the building. Revit has different features such as parametric modeling, various interoperability options, work sharing, as well as analyzing spaces. Simple user interface and easy to understand tutorial videos enable beginner designer to fully develop accurate 3D model of the buildings.

2.4.2 EnergyPlus Software

Meanwhile, EnergyPlus software is chosen as the simulation software because it is interoperable with Autodesk Revit software. Besides that, EnergyPlus is the only software that has the option to set PCM as the parameters in their building simulation. Additionally, it uses heat balance algorithm and conduction finite difference to measure the thermal and enthalpy properties during the simulation.

2.5 Interoperability of Revit with EnergyPlus Software

The IFC file format from Revit can be converted to IDF file for energy simulation by EnergyPlus using the space boundary tools (SBT-1). Since the process of transferring files from Revit to EnergyPlus is not direct, SBT is used as an interoperable bridge for exchanging data [13]. Space boundary tool is a stand-alone tool for processing validated IFC file from any BIM software. It consists of two subprocesses before changing the file to IDF format. The first process is adding space boundaries to the IFC file, followed with the second subprocess which is simplification of IFC file with space boundaries for the use of BIM software such as EnergyPlus [14].

3 Methodology

3.1 Data Acquisition

The building's data acquisition processes are divided into two which are acquisition of geometrical data and energy modeling data. Geometrical data that are essential in developing the design structure of the building such as height of the building, window sizes, wall thickness and wall geometry are acquired through manual measurement using measuring tape and other related tools. After that, parameter data such as the material used in the building, window glazing, zone, material and fenestration process are needed for the energy simulation process. Figure 1 shows the floor plan of the Malaysia building.

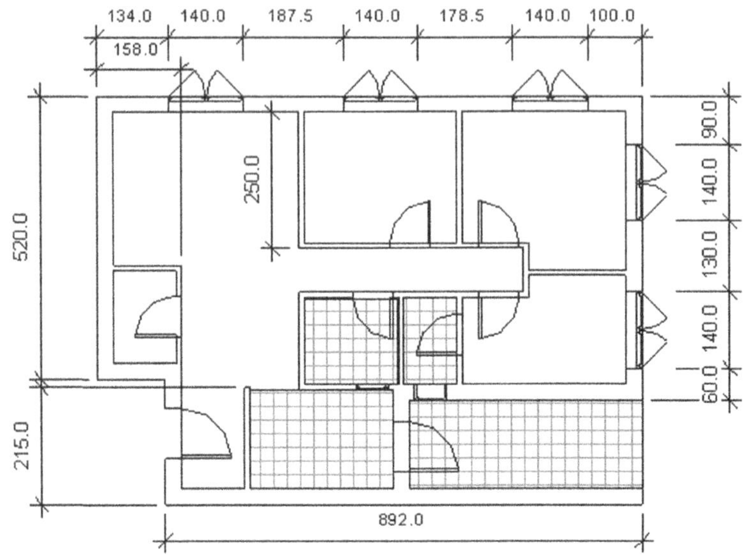

Fig. 1 Floor plan of the building

3.2 Revit Modeling

The data collected will be used into designing the building 3D model by using Autodesk Revit. From the building floor plan, the 3D model of the building is built. In order to ensure the building components is identifiable in EnergyPlus software, all the building components such as the walls, floors and windows are made from generic template that is readily available in the Revit. Figure 2 shows the final design of the building.

Before exporting the model into EnergyPlus software, the design must undergo space boundary set. The spaces later will be used as zones in the EnergyPlus. The finished 3D model is saved as IFC file and then exported to the space boundary tools (SBT-1) to be converted into IDF file format. After importing the IFC file of the design into SBT program, there are few steps before it can generate a "clean" IDF file for EnergyPlus.

First, the space boundary will be calculated based on the space created inside the 3D model. The space will have a tolerance of 0.01 m and will be calculated automatically by the program. After that, the construction materials of the building component are mapped according to its respective mapping target. For example, windows in Revit will be mapped with window component in SBT. This will ease the process of transforming the components from Revit to EnergyPlus. Finally, the generated IDF file is ready to be exported into EnergyPlus software.

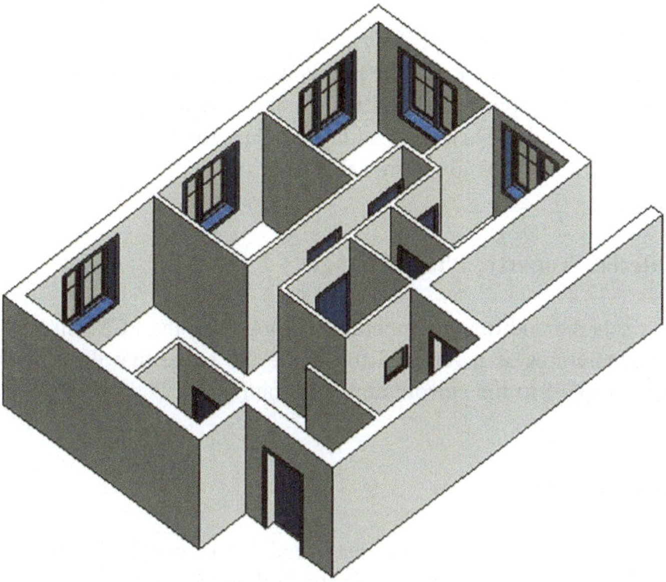

Fig. 2 Final 3D design of the building using Revit

3.3 *EnergyPlus Design and Modeling*

Before importing the IDF file, the location of the building is set in Kuching, Sarawak, Malaysia. Therefore, the meteorological data of this location are downloaded from the EnergyPlus Web site. Next, the IDF file is edited for its internal and external environment design. This step is crucial for the software to produce most accurate result for the simulation. There are a few parameters set in this step. Below are the settings of the most important parameters for the simulation.

3.3.1 Building

The name of the building is set as Tabuan Plaza Residential Building, and the location is set into suburbs because it is located in residential area. Meanwhile, the solar distribution is set into full exterior.

3.3.2 Sizing Period and Run Period

The period of both sizing and run simulation is set from 1 January until 31 December. Hence, this will produce a year worth of result.

3.3.3 Material

The materials are already set using the previous SBT program. However, in order to differentiate from the typical wall and PCM-enhanced wall, the properties for each wall are set accordingly. Therefore, additional data for both the PCM and default wall material are needed as given in Table 1.

3.3.4 Material Property: Phase Change

This section sets the additional temperature property of the PCM insulation. These ranges of temperatures as given in Table 2 are the range that the wall will experience when exposed to the surrounding environment.

3.3.5 Internal Gains

The internal gains of the building consist of number of people, lights and electrical equipment. The number of people inside the building is set as 4 people, and a total of 72 W of lighting level, while a total of 1454 W of electrical equipment with a fraction loss of 0.25 is set. These parameters are set as constant for both default wall and PCM wall simulations.

Table 1 Additional data input for PCM-enhanced cellulose and typical wall

Name	PCM-enhanced cellulose wall	Typical wall
Roughness	Medium rough	Medium rough
Thickness (m)	0.001 ~ 0.1	0.001 ~ 0.1
Conductivity (W/m K)	0.0447020014	1.95
Density (kg/m^3)	61.1904902	2240
Specific heat (J/kg K)	1481.13601	900
Thermal absorptance	0.9	0.9
Solar absorptance	0.7	0.7
Visible absorptance	0.7	0.7

Table 2 Range of temperatures and their enthalpy

Temperature (°C)	Enthalpy (J/kg)	Temperature (°C)	Enthalpy (J/kg)
0	0	32	59,762
24	35,550	34	74,876
25	37,069	36	86,512
26	38,661	37	89,043
27	40,383	38	90,712
28	42,354	39	92,209

3.3.6 HVAC Templates

To create a simple ideal load system to use for sizing and oriented simulations, HVAC Template: Thermostat and HVAC Template: Zone: Ideal Load Air System are both used in the EnergyPlus. For the thermostat, the constant heating set-point and cooling set-point are set as 15 and 25 °C, respectively, while for zone template, the most important parameter is outdoor flow rate per zone. This input is calculated by dividing the environment air flow rate inside the room with the area of the room, and hence, the value is set as 0.0153 m^3/s.

3.4 Simulation and Retrofitting

To ensure the data obtained from the simulation are accurate and dependable, the simulation was run twice. The first simulation will be without the PCM module, and the second simulation will involve PCM-enhanced cellulose encapsulated into the walls of the building. The simulation is set to run throughout the year. This means that the data of a whole year will be produced by the EnergyPlus.

3.5 Data Extraction and Comparison

The results from the EnergyPlus simulation will be compared between both tests. Energy-related output from the simulation such as the heat flux and temperature of the walls will be extracted and tabulated. The comparison of both results will be analyzed. The cooling load savings potential of the both walls is measured by measuring the average heat flux of the walls.

4 Results and Discussion

4.1 Typical Wall Simulation Results

Simulation results for the typical wall are shown in Tables 3, 4, and 5. Table 3 gives the summary on annual building utility performance for entire facility, Table 4 lists the monthly zone cooling, and Table 5 provides monthly energy consumption district cooling.

Table 3 Annual building utility performance summary for entire facility

	Total energy (kWh)	Energy per total building area (kWh/m^2)	Energy per conditioned building area (kWh/m^2)
Total site energy	30825.46	613.97	613.97
Net site energy	30825.46	613.97	613.97
Total source energy	60842.53	1211.84	1211.84
Net source energy	60842.53	1211.84	1211.84

Table 4 Zone cooling summary monthly

Month	Zone air system sensible cooling energy (kWh)
January	829.21
February	882.95
March	1040.82
April	1151.79
May	1307.07
June	1439.06
July	1356.01
August	1242.96
September	1026.21
October	967.2
November	913.8
December	906.98
Annual sum or average	13064.05
Minimum of months	829.21
Maximum of months	1439.06

4.2 PCM-Enhanced Wall Simulation Results

Tables 6, 7, and 8 show the simulation results of PCM-enhanced wall. Table 6 gives the summary on annual building utility performance for entire facility, Table 7 lists the monthly zone cooling, and Table 8 provides monthly energy consumption district cooling.

The results from the simulation are represented in charts for easier comparison between both typical wall and PCM-enhanced wall. From Fig. 3, the total site energy for typical wall is 30825.46 kWh while PCM wall is 25817.44 kWh. The total source energy are 60842.53 kWh and 55555.73 kWh for typical wall and PCM wall, respectively. This means that the typical wall in the building uses 11.23% more total energy compared to PCM-enhanced wall, annually. These values are calculated including all the parameters set inside the facility such as total number of equipment, number of people, HVAC system, as well as lighting.

Table 5 Energy consumption district cooling monthly

Month	District cooling facility (kWh)	District cooling facility {maximum} (W)	District cooling facility {TIMESTAMP}
January	1174.49	3561.22	30-JAN-17:19
February	1204.63	4021.47	13-FEB-17:49
March	1416.79	4178.55	30-MAR-17:30
April	1534.98	4367.8	11-APR-17:19
May	1704.68	5289.09	07-MAY-17:49
June	1833.84	5339.79	13-JUN-18:10
July	1732.75	5134.21	17-JUL-18:10
August	1611.14	4563.75	20-AUG-17:19
September	1385.77	3976.09	03-SEP-17:49
October	1325.38	4581.58	27-OCT-17:30
November	1250.87	4485.98	24-NOV-17:30
December	1245.75	3850.77	18-DEC-17:19
Annual sum or average	17421.07		
Minimum of months	1174.49	3561.22	
Maximum of months	1833.84	5339.79	

Table 6 Annual building utility performance summary for entire facility

	Total energy (kWh)	Energy per total building area (kWh/m^2)	Energy per conditioned building area (kWh/m^2)
Total site energy	25817.44	514.22	514.22
Net site energy	25817.44	514.22	514.22
Total source energy	55555.73	1106.54	1106.54
Net source energy	55555.73	1106.54	1106.54

Figure 4 shows the monthly zone cooling graph of both walls. Zone cooling walls measured the internal cooling potential of the building. This takes into account several parameters such as the HVAC system and internal gains. Based on the EnergyPlus simulation, the sum of annual energy for typical and PCM-enhanced wall is 13064.05 kWh and 8261.4 kWh, respectively. This means that PCM-enhanced wall uses 36.76% less energy in terms of cooling the entire building.

The EnergyPlus simulation showed that the trend of energy usage for cooling both the walls is increasing throughout the year until June and decreases smoothly until the end of December. The maximum cooling potential recorded was in June that comprised 10.8% of the total energy savings.

Table 7 Zone cooling summary monthly

Month	Zone air system sensible cooling energy (kWh)
January	540.19
February	567.74
March	665.23
April	726.3
May	802.98
June	866.67
July	815.5
August	773.2
September	658.12
October	642.17
November	609
December	594.28
Annual sum or average	8261.4
Minimum of months	540.19
Maximum of months	866.67

Table 8 Energy consumption district cooling monthly

	District cooling facility (kWh)	District cooling facility {maximum} (W)	District cooling facility {TIMESTAMP}
January	865.5	2029.19	09-JAN-18:19
February	871.94	2214.18	13-FEB-19:00
March	1022.06	2364.61	30-MAR-19:10
April	1091.42	2419.84	11-APR-19:40
May	1183.99	2700.52	07-MAY-18:40
June	1247.18	2768.94	01-JUN-18:49
July	1176.69	2639.1	17-JUL-19:19
August	1126.9	2504.88	20-AUG-18:19
September	1000.45	2258.12	03-SEP-19:00
October	981.46	2551.34	27-OCT-18:40
November	929.35	2588.85	24-NOV-18:40
December	916.12	2209.83	15-DEC-18:19
Annual sum or average	12413.06		
Minimum of months	865.5	2029.19	
Maximum of months	1247.18	2768.94	

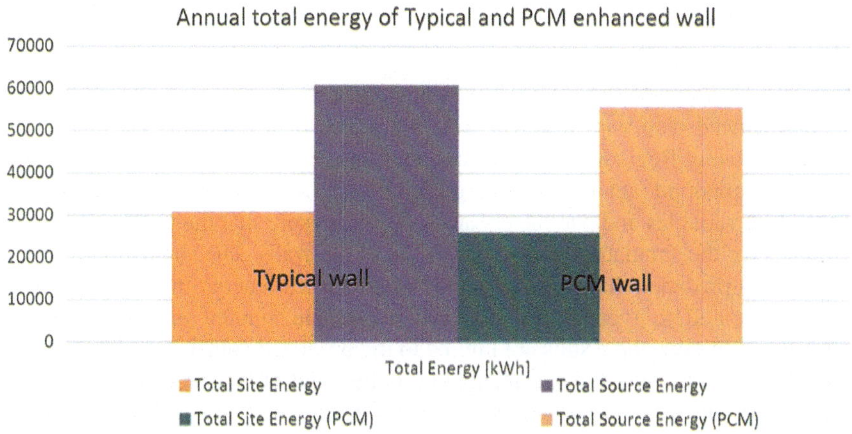

Fig. 3 Annual total energy comparison

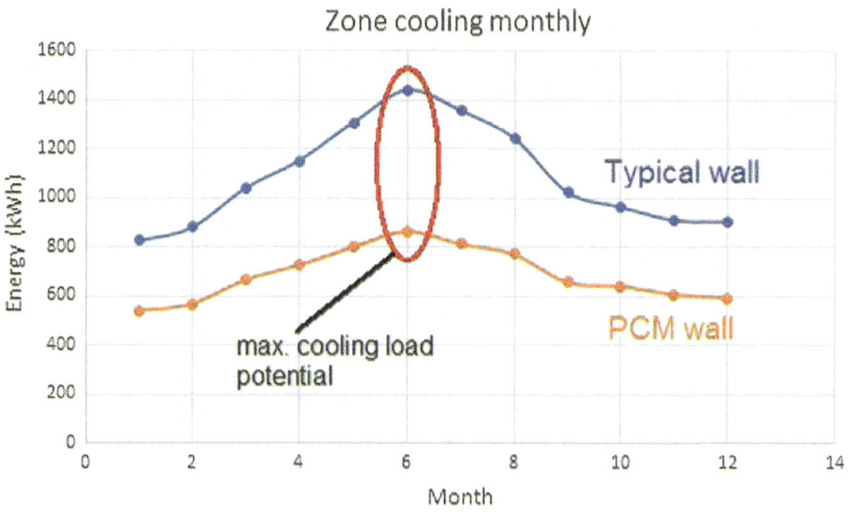

Fig. 4 Zone cooling monthly for typical and PCM-enhanced walls

A previous similar research paper [14] concluded that the PCM can reduce the energy consumption by 5% from a short period of 4 days while this research paper predicted reduction of 11.23% annually. The difference in total value may be due to different parameter inputs such as location of the building, size of the building, internal gains, running period as well as type of PCM used for the simulation.

5 Conclusions

This paper compares the EnergyPlus simulation of typical wall as well as PCM-enhanced wall with different wall properties of a building. The 3D model developed using Revit software was exported into Space Boundary Tool program and then exported again into EnergyPlus software to ensure interoperability. Controlled variables and parameters are set to make sure that the only manipulated variables for the simulation are the properties of the walls. The simulation from the EnergyPlus predicted that PCM-enhanced wall uses less energy in terms of cooling when compared to typical wall. It gave reasonable amount of energy savings annually. However, the results are limited to the predicted values only. Therefore, a paper based on the actual experimentation on the field of the same building can be further implemented to produce more accurate results and comparison.

References

1. Pérez-Lombard L, Ortiz J, Pout C (2008) A review on buildings energy consumption information. Energy Build 40(3):394–398
2. Ma Z, Cooper P, Daly D, Ledo L (2012) Existing building retrofits: methodology and state-of-the-art. Energy Build 55:889–902
3. Chan A (2011) Energy and environmental performance of building façades integrated with phase change material in subtropical Hong Kong. Energy Build 43(10):2947–2955
4. Rogers J, Preece C, Chong HY (2015) Adoption of Building Information Modelling technology (BIM): perspectives from Malaysian engineering consulting services firms. Eng Constr Archit Manag 22(4):424–445
5. Habibi S (2017) The promise of BIM for improving building performance. Energy Build 153:525–548
6. Krarti M (2008) Energy efficient systems and strategies for heating, ventilating, and air conditioning (HVAC) of buildings. J Green Build 3(1):44–55
7. Agarwal Y, Balaji B, Dutta S, Gupta RK, Weng T (2011) Duty-cycling buildings aggressively: the next frontier in HVAC control. In: Proceedings of the 10th ACM/IEEE international conference on information processing in sensor networks, pp 246–257
8. Rao VV, Parameshwaran R, Ram VV (2018) PCM-mortar based construction materials for energy efficient buildings: a review on research trends. Energy Build 158:95–122
9. Dincer I, Colpan C, Kizilkan O, Ezan M (2015) Progress in clean Energy: volume 2, novel systems and applications (1st ed 2015). https://doi.org/10.1007/978-3-319-17031-2
10. Tenorio J, Sanchez-Ramos J, Ruiz-Pardo A, Alvarez S, Cabeza L (2015) Energy efficiency indicators for assessing construction systems storing renewable energy: application to phase change material-bearing facades. Energies 8(8):8630–8649
11. Sanhudo L, Ramos NMM, Poças Martins J, Almeida RMSF, Barreira E, Simões ML, Cardoso V (2018) Building information modeling for energy retrofitting—a review. Renew Sustain Energy Rev 89:249–260
12. Chiaia B, Davardoust S, Osello A, Aste N, Mazzon M (2014) BIM and interoperability for energy simulations. International Building Performance Simulation Association, 28 Nov 2019. http://www.ibpsa.org/proceedings/BSA2015/9788860460745_13.pdf

13. Lab B (2018) IFC Space Boundary Tool (SBT), Department of Energy National Laboratory, University of California, 28 Nov 2019. https://gaia.lbl.gov/interoperability/SBT/
14. Shrestha S, Miller W, Stovall TK, Omer Desjarlais A, Childs K, Porter W, Bhandari M, Coley SJ (2011) Modeling PCM-Enhanced Insulation System and Benchmarking EnergyPlus against Controlled Field

Shade Net to Reduce Building Cooling Load: An Experimental Study with RCC and GI Sheet Roofs

Vijesh V. Joshi

Abstract Use of shade net over the roofs could be one of the economic ways to reduce heat gained by the buildings in arid regions. Model rooms with reinforced cement concrete (RCC) and galvanized iron (GI) sheet roofs were set up in the open field. 50% shading nets were tried over the roofs of the model rooms at a height of 125 mm from the roof top surface. From the experiments, it is concluded that using shade net over GI sheet roof can be as good as a RCC roof without shade net. These shade nets are found to cause negligible adverse effects on the night time cooling. Around 8 and 16 °C reduction in peak temperatures of roof surfaces due to shade net were observed in RCC and GI sheet roof cases, respectively. In turn, 1 and 2 °C reduction in average inside air temperatures has been observed in the respective cases.

Keywords Building heat transfer · Energy efficiency in buildings · Passive building cooling techniques · Thermal comfort · Built environment

1 Introduction

An immediate action is required to reduce energy consumption in buildings due to heating, ventilation, and air conditioning (HVAC) devices. HVAC devices have had been accounting to about 30–40% of the building energy consumption [1, 2]. In either warm tropics or hot arid regions, building heating (solar gain) is an unwelcoming process and hence needs a proper treatment for the buildings to be energy efficient. On the other hand, earlier studies have shown that up to 60% of heat transfer occurs through the roof [3–5], and hence, it is the most important element to target to reduce the solar gain.

V. V. Joshi (✉)
School of Mechanical Engineering, Vellore Institute of Technology, Vellore, India
e-mail: vijesh.joshi@vit.ac.in; vijeshjoshi23@gmail.com

Mechanical Engineering, Indian Institute of Science, Bangalore, India

K. Thirumaran et al. (eds.), *Sustainable Urban Architecture*,
Lecture Notes in Civil Engineering 114,
https://doi.org/10.1007/978-981-15-9585-1_14

Many passive cooling techniques are being used to minimize the heating of the built environment due to solar radiation [6]. Use of cool roofs is found to effectively reduce the solar gain but the efficacy decreases with the usage [7]. On the other hand, the idea of roof gardens and roof ponds can be implemented in places where water is available abundantly. In hot arid/warm tropic regions (with water scarcity), we essentially need some economic and effective ways to reduce solar gain of buildings. Use of shade net could be one such simple technique and is the subject of the present paper. Shade nets are generally used in green houses to have reduced solar radiation but allowing air exchange depending on the plants' requirement. Though a lot of literature is available related to shade nets like [8, 9], the studies were carried out with the view of using shade nets for green houses. However, there exists a lack in the study of the effect of shade net on solar gain in the case of buildings. According to 2011 census, in India, 29% of the buildings have RCC roof and 15.9% have GI sheet/metal/asbestos roofs [10] (i.e., around 45% of the buildings in India do have either of the two roof types). Hence, we have studied the effect of shade net on the inside air temperature of two model rooms: one with RCC roof and the other with GI sheet roof. The shade net is cheaper and can be immediately implemented in the case of either already existing buildings or new buildings.

2 Experimental Set Up Details

The dimensions of the model rooms studied are 1 m × 1 m × 0.3 m in size as shown in Fig. 1. The side walls of the model rooms were essentially adiabatic in nature as they were made of 20-cm-thick thermocol and outer surface of the bottom RCC slabs was well insulated. This arrangement allowed heat transfer between inside and outside the enclosure only through the roof. The shade net used in the experiments has the quality of reducing solar radiation by 50%. The shade was ensured to remain at and parallel to the roof surface at a distance of 125 mm (6 in.) using a frame as shown in Fig. 2. For every set of two days, the conditions of experiments were changed by either keeping or removing shade net from over the roof surface (refer Table 1). Sufficient additional length of the shade net was used in order to ensure that the roof surface was always under the shade. This configuration allows us to quantitatively comment on the temperature reductions due to the shade net over the roofs.

In the case of model room with RCC roof, the first four days the roof was not covered with shade net and the next four days it was covered with shade net. On the other hand, the model room with GI sheet roof had shade net over the roof from third to sixth day of eight days (refer Table 1). The last two days (seventh and eighth day) were cloudy. However, the maximum solar radiation was around 850 W/m^2–900 W/m^2 and the maximum dry bulb temperature was in the range of 34–36 °C. For the details of abbreviations used, refer Fig. 1.

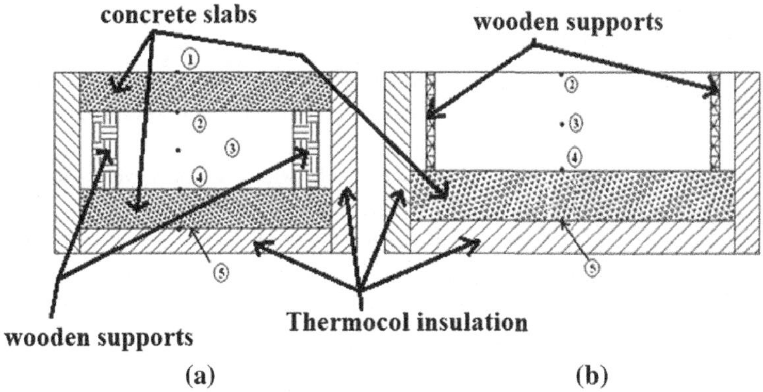

Fig. 1 **a** Model room with RCC roof and **b** Model room with GI sheet roof. (Dots in the figure indicate the location of thermocouples). 1-Top slab top surface (TSTS); 2-Top slab bottom surface (TSBS); 3-Enclosure air (ENCL AIR); 4-Botoom slab top surface (BSTS); 5-Bottom slab bottom surface (BSBS)

Fig. 2 Shade net used in the present experiments. Studs and nuts were used in the corners to ensure the horizontal orientation of the shade net. Two such configurations were set up and were kept over the RCC and GI sheet roofs in different combinations as mentioned in Table 1

Table 1 Details of the experiments planned during the month of April 2013 (typical days of summer weather in Bangalore, India)

Dates	Shading conditions
April 10–11	Both GI sheet and RCC roofs—without shade net covering
April 12–13	GI sheet roof—with shade net covering RCC roof without shade net covering
April 14–15	Both GI sheet and RCC roofs—with shade net covering
April 16–17 (rained on these two days for some time)	GI sheet roof—without shade net covering RCC roof with shade net covering

T-type thermocouples were used for the temperature measurements. In the case of GI sheet roof, due to high thermal conductivity and low thermal mass of GI sheet, both inner and outer surface temperatures are assumed to be same. Hence, only the inner surface temperature T_{TSBS} was measured. A pyranometer was used for the solar radiation measurement and a thermohygrometer was used for the temperature and humidity measurements. Wind speed and rain fall were recorded using a wind anemometer and a tipping bucket rain gauge. The maximum wind speed during the experiments was around 3.5 m/sec–4 m/sec. The accuracy in case of the pre-calibrated wind anemometer was ±0.89 m/sec with starting threshold of 1.33 m/sec and the sampling interval was 11 s. Agilent data logger was used for recording thermocouple temperatures and the solar radiation measurements. A TFA data logger was used to record the weather parameters. The variables were recorded for every 10 min. However, it was ensured that the model rooms were subjected to same weather conditions.

3 Results and Discussion

Solar radiation, dry bulb, and wet bulb temperatures as observed on the days of experiments conducted are as shown in Fig. 3. The experiments were conducted in the month of April. Bangalore city experiences typical tropical summer, and in the recent years, the city has recorded an increase in the urban air temperature followed by an obvious increase in the cooling load of buildings during summer. Hence, the experiments were conducted under summer weather conditions. Temperature variations in the case of model room with RCC roof and GI sheet roof are as shown in Figs. 4 and 5. When the shade net was placed over TSTS, a sudden drop in T_{TSTS} was observed in the case of both model rooms. The sudden drop (in less than ten minutes after the shade net was placed over the roof) is 0.8 °C in the case of RCC roof and 7 °C in the case of GI sheet roof outer surfaces, respectively.

Figure 4 shows the temperature variations in the case of model room with RCC roof: first day is without shade net, and the next day is with shade net. Peak

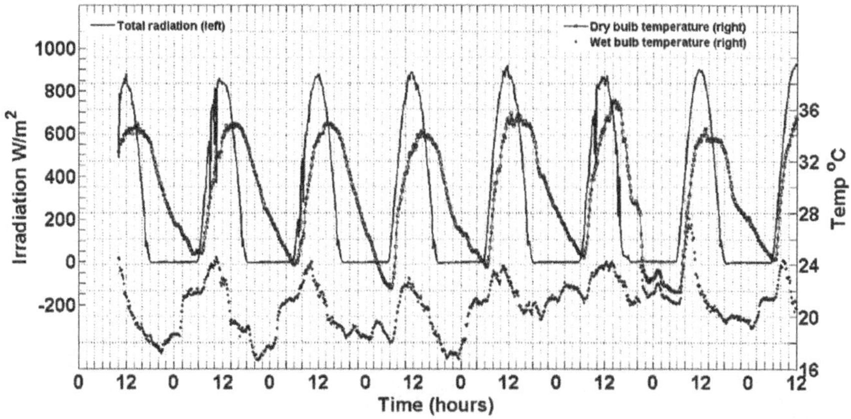

Fig. 3 Solar radiation, dry bulb, and wet bulb temperature (10–18 April 2013). Y-axis on the left side gives the solar irradiation and on the right side, the temperatures (both dry bulb and wet bulb)

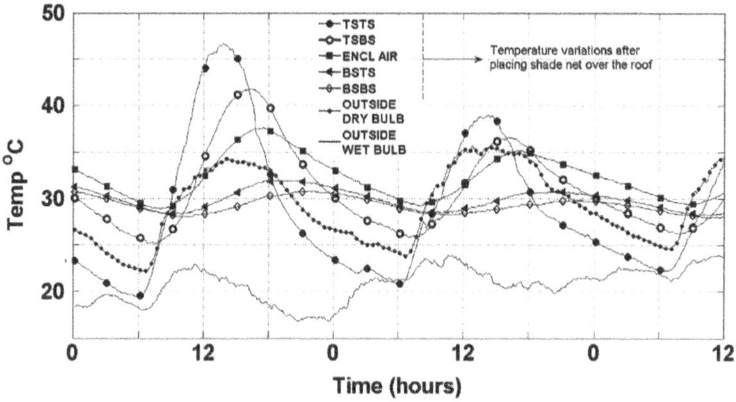

Fig. 4 Temperature variation with time—model room with RCC roof (with shade net and without shade net)

temperature of top slab top surface dropped down by 8 °C by using shade net. Further, T_{TSBS}, $T_{ENCLAIR}$, and T_{BSTS} also dropped by 5.5, 2.5, and 1.5 °C, respectively.

Similarly, in the case of model room with GI sheet roof, we observed reduction in peak temperatures of T_{TSTS}, $T_{ENCLAIR}$, and T_{BSTS} by 16, 6, and 3 °C, respectively. Tables 2 and 3 give the maximum, minimum, and average temperature values with and without shade net. Enclosure air temperature averaged over the day is found to be reduced by about 1 °C in the case of RCC roof and the same is about 2 °C in the case of GI sheet roof. This is important as far as energy efficiency is concerned.

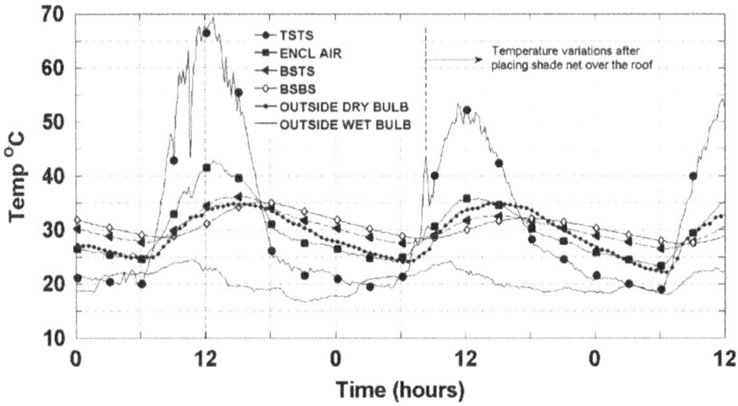

Fig. 5 Temperature variation with time—model room with GI sheet roof (with shade net and without shade net)

Table 2 Maximum, minimum, and average of temperatures of concrete slabs and enclosure air—RCC roof case

RCC roof	T_{TSTS}		T_{TSBS}		$T_{ENCL\ AIR}$		T_{BSTS}		T_{dry}		T_{wet}	
	A[a]	B[b]	A	B	A	B	A	B	A	B	A	B
MAX	46.7	39	41.8	36.6	37.6	35.2	32	30.7	34.3	35.7	23	24
MIN	20.7	22.1	25.8	26.3	29.2	29.4	28.5	28.3	23.9	24.8	16.8	19.9
AVG	30.5	28.9	32.5	30.8	33.1	32.2	30.3	29.7	29.3	30.5	15.6	17.9

[a]A-without shade net (April 13, 2013)
[b]B-with shade net (April 14, 2013)

Table 3 Maximum, minimum, and average of temperatures of concrete slabs and enclosure air—GI sheet roof case

GI sheet roof	T_{TSBS}		$T_{ENCL\ AIR}$		T_{BSTS}		T_{dry}		T_{wet}	
	A[a]	B[b]	A	B	A	B	A	B	A	B
Max	69.5	53.4	42.8	35.9	36.3	32.6	34.8	34.8	24.5	24.2
Min	19.1	18.6	24.2	23.3	27.6	26.4	24	22.1	16.8	16.9
Avg	34.3	31	31	29	31.5	29.7	29.8	29.3	15.3	15.2

[a]A-without shade net (April 11, 2013)
[b]B-with shade net (April 12, 2013)

Comparison of temperature variations on the days when the GI sheet roof was under the shade net are shown in Fig. 6. From the day the shade net was placed over the roof, in the case of GI sheet roof, the first three days sky was clear. Unlike the sudden temperature drop when the shade net was put over the GI sheet roof (refer Fig. 5—second day around 8:30 a.m.), smooth variation in the roof surface temperature was observed during second and third consecutive days. Similarly,

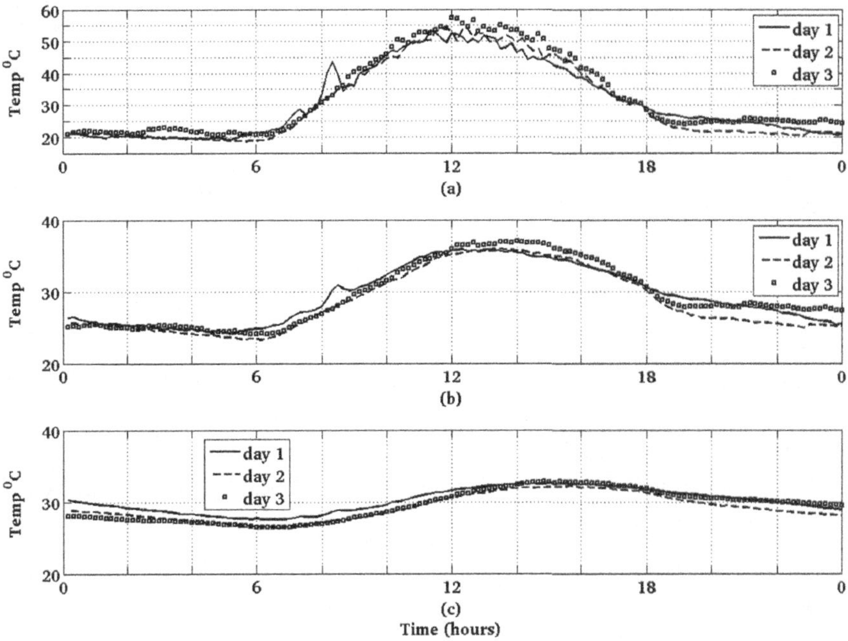

Fig. 6 Temperature variation with time: first, second, and third day after the shade net was put over model room with GI sheet roof. **a** Top slab top surface, **b** enclosure air, and **c** bottom slab top surface

enclosure air temperature and bottom slab top surface temperature were found to have smooth modulations on second and third days, respectively. On the other hand, the maximum and minimum temperatures attained during the three days after the shade net was put, the modulations seem to repeat. Hence, the effects of shade net over GI sheet roof can be considered to get stabilized in few hours of time; low thermal mass of the GI sheet is the primary reason relatively early stabilization as compared to RCC roof case.

Referring to Fig. 7, in the case of RCC roof, the response of the model room was slow due to high thermal mass of the roof. The temperatures were reducing from day to day. On the second day after the shade net was placed over the RCC roof, in the night at around 8:30 p.m., there was light rain for few minutes. The temperatures on third day will have the effect of rain. But, on the fourth day, if we look at temperatures in the evening time, they are close to the second-day data. No raining would have resulted in similar modulation on second and third day; however, the second-day temperatures are lower than that on first day. Hence, we can say that the thermal stabilization in the case of RCC roof takes around 24 h.

The maximum and minimum of roof surface temperatures in case of model room with GI sheet roof "with shade net" (refer Table 2) are around 53.4 and 18.6 °C, respectively; average temperature (over 24 h) is 31 °C. While the same in case of

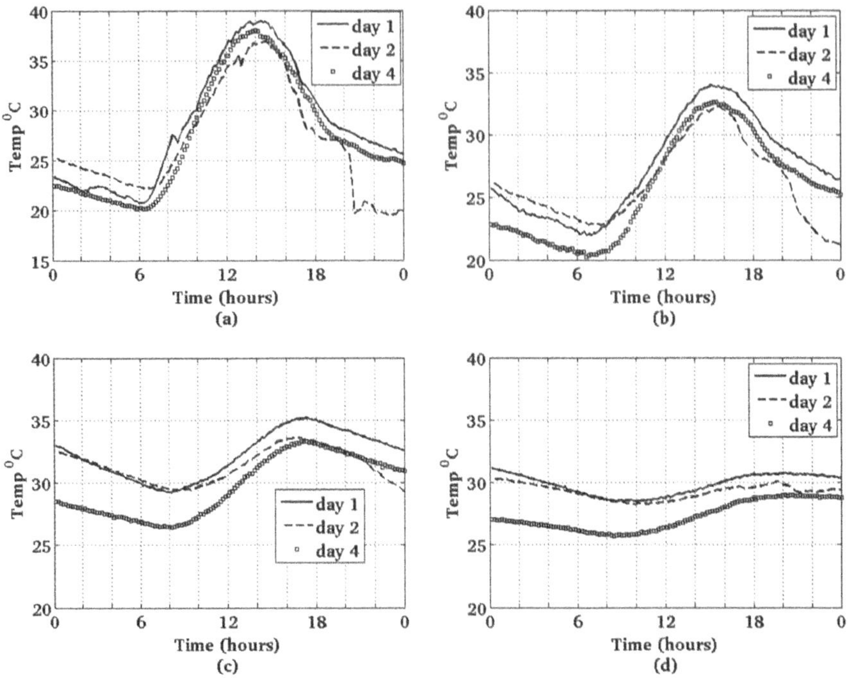

Fig. 7 Temperature variation with time: first, second, and fourth day after the shade net was put over model room with RCC roof. **a** Top slab top surface, **b** top slab bottom surface, **c** enclosure air, and **d** bottom slab top surface

RCC roof "without shade net" are 46.7 and 20.7 °C, respectively; average temperature (over 24 h) is 30.5 °C. The shade net over either of the roofs directly affects the roof surface temperatures. Looking at the average temperatures of the roofs, it can be said that GI sheet roof with 50% shade net could be as effective as RCC roofs without shade net. From the average enclosure air temperatures and by assuming the enclosure air to be well mixed, reduction in heat gain was determined. The percentage of reduction in heat gained by the enclosure air by the use of shade net is found to be 3.4 and 6.1% in the case of model rooms with RCC roof and GI sheet roof, respectively, with negligible adverse effects in the night and during early morning. The presence of shade net seems to have negligible effects on night radiation cooling of the roofs.

Tables 4 and 5 give the details of number of hours during which T_{TSBS} and $T_{ENCLAIR}$ are in particular ranges for both with and without shade net cases. Considerable decrease in number of hours of heat gain can be observed with the use of shade net in both the cases. In the case of RCC roof, top slab bottom surface was found to be above 35 °C for about 6 h without shade net in a given 24 h cycle. But, with shade net, it was always lower than 35 °C (Table 4). On the other hand, enclosure air was above 35 °C for about 15 h. But with shade net, it was always

Table 4 Number of hours of T_{TSBS} under different temperature ranges in a 24 h cycle

Roof type	T_{TSBS}											
	20–25 °C		25–30 °C		30–35 °C		35–40 °C		40–45 °C		>45 °C	
	A[a]	B[b]	A	B	A	B	A	B	A	B	A	B
RCC	0	0	11	11	5	9	4	0	4	0	0	0
GI sheet	13	11	0.5	2.5	1.5	2	0.5	1	1.5	2	7	5.5

[a]A-without shade net
[b]B-with shade net

Table 5 Number of hours of $T_{ENCLAIR}$ under different temperature ranges in a 24 h cycle

Roof type	$T_{ENCLAIR}$											
	20–25 °C		25–30 °C		30–35 °C		35–40 °C		40–45 °C		>45 °C	
	A[a]	B[b]	A	B	A	B	A	B	A	B	A	B
RCC	0	0	4	5	12.5	19	17.5	0	0	0	0	0
GI sheet	0	5	13.5	10	2	6.5	5	2.5	3.5	0	0	0

[a]A-without shade net
[b]B-with shade net

below 35 °C. In the case of GI sheet roof also, there were similar reductions in the number of hours with higher temperatures.

The model rooms essentially have 20-cm-thick thermocol (typical thermal conductivity—0.03 W/m K) as side walls and bottom concrete slab was also insulated from outside with 10-cm-thick thermocol. Though there could be heat transfer possible through these thermocol walls, it would be negligible as compared to either concrete (typical thermal conductivity—1.2 W/m K) or GI sheet (typical thermal conductivity—60 W/m K). Hence, the author assumes that the effect of heat transfer either through side walls or from the bottom concrete slab to the outside surroundings is negligible. Therefore, the analysis carried out helps us to understand the role of shade net over the RCC roof in reducing the inside air temperature of an enclosed space. However, the analysis is limited to typical tropical summer weather conditions for shallow enclosure air spaces.

4 Conclusion

Use of shade net over the roofs is found to have significant potential to reduce heat gain. On the other hand, it is inexpensive and can be implemented immediately in arid and warm tropical regions. The presented results are for the case of shade net which cuts down 50% solar radiation. Considerable decrease in temperatures by the use of shade net was observed. In the case of RCC, the top slab bottom surface temperature was found to be lower than 40 °C with shade net throughout a 24 h cycle but without shade net, temperature was above 40 °C for around 4 h. On the other hand, the enclosure air temperature was observed to be below 35 °C throughout 24 h cycle with shade net, and without shade net, the temperature was more than 35 °C for around 7.5 h. Similarly, significant temperature reductions with shade net in the case of GI sheet have been observed. Using shade net over GI sheet roof can make it as good as the bare RCC roof as far as inside air temperature is concerned. Around 3.4 and 6.1% reduction in heat gain in the case of RCC roof and GI sheet roof, respectively, in a 24 h cycle could be achieved simply by covering the roof by shade net (50% shading net). It is also observed that the night time and early morning temperatures are not much affected due to the presence of shade net. That is, the heat loss from the roof outer surface during night time and early morning remained same with and without the shade net. As the percentage of buildings with RCC and GI sheet roofs are significant in most of the warm tropic regions, use of shade net over the roofs can considerably reduce the HVAC energy demand. The work may be extended to see the effects with shade nets of different sizes and at different heights from the roof surface. The shade nets can also be experimented over different roof types.

Acknowledgements This research was funded in part by Center for infrastructure, Sustainable Transportation and Urban Planning (CiSTUP), IISc. The author is immensely grateful to Jaywant H. Arakeri (Professor, IISc-Bangalore) and K. R. Sreenivas (Professor, JNCASR-Bangalore) for their guidance. Author is solely responsible for any errors if found and should not tarnish the reputations of these esteemed persons.

References

1. DOE BEDB (2011) Chapter: 1 Buildings Sector. 1.1 Buildings Sector Energy Consumption. 1.1. 1 US Residential and Commercial Buildings Total Primary Energy Consumption. US Department of Energy (DOE)—Energy Efficiency and Renewable Energy, 6
2. Outlook AE (2010) Energy information administration. Department of Energy 92010(9):1–15
3. Amer EH (2006) Passive options for solar cooling of buildings in arid areas. Energy 31(8–9):1332–1344
4. Maneewan S, Hirunlabh J, Khedari J, Zeghmati B, Teekasap S (2005) Heat gain reduction by means of thermoelectric roof solar collector. Sol Energy 78(4):495–503
5. Soubdhan T, Feuillard T, Bade F (2005) Experimental evaluation of insulation material in roofing system under tropical climate. Solar Energy 79(3): 311–320
6. Geetha NB, Velraj R (2012) Passive cooling methods for energy efficient buildings with and without thermal energy storage—A review. Energy Educ Sci Technol Part A: Energy Sci Res 29(2):913–946
7. Elie Tawil PE (2010) Guidelines for selecting cool roofs
8. Cohen S, Fuchs M (1999) Measuring and predicting radiometric properties of reflective shade nets and thermal screens. J Agric Eng Res 73(3):245–255
9. Ilić ZS, Milenković L, Stanojević L, Cvetković D, Fallik E (2012) Effects of the modification of light intensity by color shade nets on yield and quality of tomato fruits. Sci Hortic 139:90–95
10. CO_GEN NGB (2007) Bureau of energy efficiency

The manufacturer's authorised representative in the EU is Springer Nature Customer Service Centre GmbH, Europaplatz 3, 69115 Heidelberg, Germany. If you have any concerns regarding our products, please contact ProductSafety@springernature.com

Printed and bound by CPI Group (UK) Ltd, Croydon, CR0 4YY

30/09/2026

02225332-0004